Basic Math
Quick Reference
HANDBOOK

Peter J. Mitas

ISBN: 978-0-615-27390-7

Cover design: Rachael Mitas (rachaelmitas.com)

Please visit
www.QRefHandbooks.com
to view and print practice problems

Contents

Whole Numbers 1

Natural Numbers and Whole Numbers 1

Digits and Numerals 1

Place Value Notation 2

Rounding a Whole Number 3

Operations and Operators 4

Adding Whole Numbers 5

Subtracting Whole Numbers 9

Multiplying Whole Numbers 12

Dividing Whole Numbers 16

Divisibility Rules 21

Multiples of a Whole Number 22

Factors of a Whole Number 24

Prime and Composite Numbers 26

Fractions 28

Fraction, Numerator, and Denominator 28

Fractions as Division 29

Proper & Improper Fractions, Mixed Numbers 30

Fractional Equivalence and Conversions 31

Equal fractions 31

Writing a fraction in lowest terms 33

Writing a fraction in higher terms 34

Equivalent improper fractions, mixed numbers & whole numbers 35

Writing an improper fraction as an equivalent whole number or mixed number........... 35

Writing a whole number as an equivalent improper fraction.. 36

Writing a mixed number as an equivalent improper fraction.. 37

Replacing several fractions with equal fractions having their least common denominator.. 38

Ordering Fractions... 39

Adding Numbers That Include Fractions........................ 40

Adding fractions that have a common denominator.. 40

Adding numbers that include fractions, whole numbers, and mixed numbers..................... 41

Adding fractions that do not have a common denominator... 42

Writing the answer to an addition problem in simplest terms... 43

Subtracting Numbers That Include Fractions................ 46

Subtracting fractions that have a common denominator.. 46

Subtracting numbers that include fractions, whole numbers, and mixed numbers..................... 47

Subtracting fractions that do not have a common denominator... 48

Subtracting fractions when the fraction to be subtracted is greater than the fraction above it.. 49

Writing the answer to a subtraction problem in simplest terms... 51

Multiplying Numbers That Include Fractions................. 52

Multiplying fractions.. 52

Multiplying numbers that include fractions and whole numbers.. 53

Multiplying numbers that include fractions and mixed numbers ... 54

Writing the answer to a multiplication problem in simplest terms ... 55

Canceling ... 56

Reciprocal fractions ... 57

Dividing Numbers That Include Fractions ... 58

Multiplying instead of dividing ... 58

Why multiply when dividing fractions ... 60

Complex fractions ... 61

Comparing Operations on Fractions ... 62

Decimal Numbers **63**

Representing Wholes and Parts as Decimal Numbers ... 63

Rounding a Decimal Number to a Specific Place Value ... 65

Ordering Decimal Numbers ... 66

Types of Decimal Numbers ... 67

Adding Decimal Numbers ... 68

Subtracting Decimal Numbers ... 69

Multiplying Decimal Numbers ... 70

Dividing Decimal Numbers ... 72

Decimal Numbers and Equivalent Fractions ... 74

Converting a fraction to a decimal number ... 74

Converting a mixed number to a decimal number ... 75

Converting a terminating decimal number to a fraction ... 75

Converting a repeating decimal number to a fraction ... 76

Ordering fractions and decimal numbers ... 76

Universal Number Concepts 77

The topics in this section describe concepts that apply to all numbers, including whole numbers, fractions, decimal numbers, and positive and negative numbers.

Set Notation 77

Powers 78

Powers of a number 78

Squares, square roots, cubes, and cube roots 80

Multiplying by powers of 10 83

Dividing by powers of 10 84

Exponential Notation 85

Exponentials 85

Simplifying exponential products and quotients 86

Writing prime factors in exponential notation 88

Multiplying and dividing by powers of 10 written in exponential notation 88

Scientific notation (for large numbers) 89

An exponent of 0 90

Fractional exponents 91

Powers of fractions and decimal numbers 92

Order of Operations 94

Properties of Numeric Operations 97

Commutative property 97

Associative property 98

Distributive property 99

Inequality Symbols 101

Graphing Numbers on a Number Line 102

Ratios, Proportions, and Percents **104**

Ratio 104

Proportion 105

Percent 106

What percent means 106

Percents and equivalent decimal numbers 107

Percents and equivalent fractions 108

Solving percent problems 109

Percents greater than 100% 112

Probability and Statistics (Selected Topics) **113**

Probability of an Event 113

Statistical Measures 114

Geometry and Measurement (Selected Topics) **117**

Angles and Lines 117

Line segment 117

Angles 117

Perpendicular and parallel lines 119

Polygons 119

Planes and polygons 119

Quadrilaterals 120

Triangles 121

The Pythagorean Theorem 121

Circles 122

Length 124

Perimeter 125

Area 126

Surface area 128

Volume 129

Positive and Negative Numbers **130**

Signed Numbers and the Number Line 130

Sign and Absolute Value 132

Opposites 134

Integers, Rational and Irrational Numbers, and Real Numbers 136

Ordering Positive and Negative Numbers 138

Adding Two Signed Numbers....................................... 140

Subtracting One Signed Number From Another 142

Expert Strategies for Adding and Subtracting Signed Numbers 144

Multiplying Signed Numbers... 146

Dividing One Signed Number by Another 149

Commutative, Associative, and Distributive Properties of Operations on Signed Numbers.............. 151

Order of Operations.. 153

Exponentials with Negative Bases 155

Exponentials with Negative Exponents 156

Multiplying and Dividing by Powers of 10 That Have Negative Exponents.. 158

Scientific Notation (for Small Numbers) 159

Using Exponentials to Identify Place Values 160

Algebra (Selected Topics) **161**

Algebraic Expressions and Equations.......................... 161

Finding the Value of an Expression.............................. 162

General Guidelines for Solving All Equations............... 163

Guidelines for Solving Equations That Have Several Operations.. 164

Begin by moving isolated constants to the side that has no variable..................................... 164

Leave a positive variable in the result 165

Remove variables from the denominator............. 166

Eliminate parentheses 167
Combine similar terms 168
Eliminate common factors 169
Solving a System of Equations 170
Using Algebra to Solve Percent Problems 171
Multiplying Monomials and Polynomials 172
Graphing on a Coordinate Grid 173
Graphing points on a coordinate grid 173
Graphing a linear equation 174
Graphing a quadratic equation 175
Formulas 176
Geometry formulas 176
The quadratic formula 177
Factorials 178
Combinations 179
Permutations 180
Arithmetic progressions 181
Geometric progressions 182
Index **183**

Whole Numbers

Natural number. Counting number. Whole number. Digit. Numeral. Place value notation. Empty place value. Expanded form. Rounding. Operation. Operator. Sum. Carry. Difference. Borrow. Check answer to a subtraction problem. Times table. Product. Quotient. Dividend. Divisor. Remainder. Check answer to a division problem. Short division. Long division. Approximately symbol (≈). Multiple. Common multiple. Least common multiple (LCM). Factor. Common factor. Greatest common factor (GCF). Divisibility rules. Prime number. Composite number. Prime factor tree.

Natural Numbers and Whole Numbers

The numbers that we use to count are called NATURAL NUMBERS or COUNTING NUMBERS.

- 18, 43, and 2 are natural numbers.
- Since we start counting with "1", zero is not a natural number.

The WHOLE NUMBERS include all the natural numbers and ZERO.

- 0, 18, 43, and 2 are whole numbers.

Digits and Numerals

The first ten whole numbers: 0, 1, 2, 3, 4, 5, 6, 7, 8, and 9 are called DIGITS. These digits are used to write every number.

Written whole numbers are sometimes called NUMERALS.

Place Value Notation

PLACE VALUE NOTATION allows the position of a digit in a number to determine its value. Each place for a digit in a whole number represents ten times the place to its right.

THOUSANDS 10 Hundreds	HUNDREDS 10 Tens	TENS 10 Ones	ONES

- 452 = 4 Hundreds + 5 Tens + 2 Ones
- 1,245 = 1 thousand+2 Hundreds+4 Tens+5 Ones

The digit "0" (zero) is used to fill EMPTY PLACE VALUES.

- 570 has no Ones.
- 507 has no Tens.
- 57 has no Hundreds, but does not need a placeholder.

COMMAS are used in numerals to separate groups of three digits.

- 32,000 32 thousand
- 25,000,000 25 million
- 10,987,654,000 10 billion, 987 million, 654 thousand

A number is written in EXPANDED FORM when it is represented as the sum of the value of each of its digits.

- 2,405 represents: 2 thousands, 4 hundreds, no tens, and 5 ones.
  In expanded form, 2,405 is written as:
  $(2 \times 1000) + (4 \times 100) + (0 \times 10) + (5 \times 1)$

Rounding a Whole Number

ROUNDING a whole number to a place value identifies which value it is closest to.

A whole number can be rounded to any specified place value.

Steps

1. Begin by locating the digit in the place value to round to.
2. Focus on the digit to its right.

- IF THE DIGIT TO ITS RIGHT IS LESS THAN FIVE replace that digit with a zero and then replace every digit to its right with a zero.

    - To round 347 to the nearest hundred, note that 3 is in the hundreds place. Since 4 is in the place to its right, and 4 is less than 5, replace the 4 with a 0 and the 7 with 0. This results in an answer of 300.

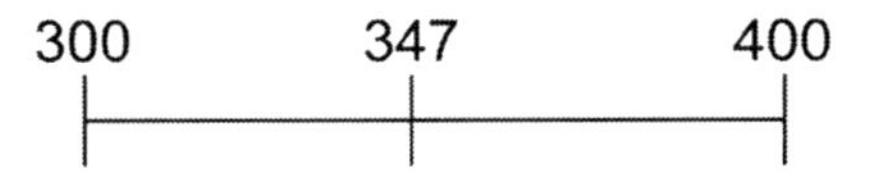

    - To round 347 to the nearest thousand, note that no number (which is really 0) is in the thousands place. Since 3 is in the place to its right, and 3 is less than 5, replace the 3 with a 0 and the 4 and 7 with 0s. This results in an answer of 000 or just 0.

- IF THE DIGIT TO ITS RIGHT IS FIVE OR GREATER replace the digit that is in the place value you are rounding to with the next higher digit, and then replace every digit to its right with a zero.

  - To round 347 to the nearest ten, note that 4 is in the tens place. Since 7 is in the place to its right, and 7 is greater than 5, replace the 4 with a 5 and the 7 with 0. This results in an answer of 350.

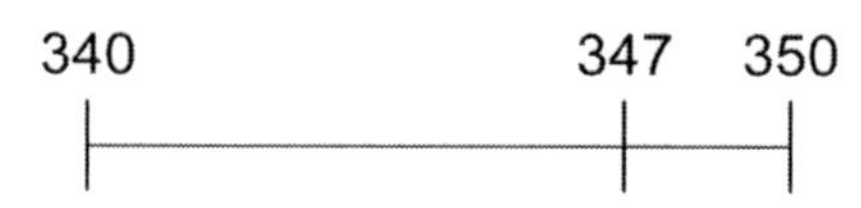

Operations and Operators

OPERATIONS are the common calculations performed on numbers: addition, subtraction, multiplication, and division.

The symbols used to indicate an operation are called OPERATORS: +, –, x, and ÷.

Placed between two numbers, an operator identifies the operation to be performed on those numbers.

Operation	Operation with Operator
Addition	$12 + 3 = 15$
Subtraction	$12 - 3 = 9$
Multiplication	$12 \times 3 = 36$
Division	$12 \div 3 = 4$

Adding Whole Numbers

Tip: In order to effectively add numbers, you should be able to immediately respond with the answer to each of the following addition facts.

$1 + 1 = 2$

$2 + 1 = 3$

$3 + 1 = 4$
$2 + 2 = 4$

$4 + 1 = 5$
$3 + 2 = 5$

$5 + 1 = 6$
$4 + 2 = 6$
$3 + 3 = 6$

$6 + 1 = 7$
$5 + 2 = 7$
$4 + 3 = 7$

$7 + 1 = 8$
$6 + 2 = 8$
$5 + 3 = 8$
$4 + 4 = 8$

$8 + 1 = 9$
$7 + 2 = 9$
$6 + 3 = 9$
$5 + 4 = 9$

$9 + 1 = 10$
$8 + 2 = 10$
$7 + 3 = 10$
$6 + 4 = 10$
$5 + 5 = 10$

$10 + 1 = 11$
$9 + 2 = 11$
$8 + 3 = 11$
$7 + 4 = 11$
$6 + 5 = 11$

$11 + 1 = 12$
$10 + 2 = 12$
$9 + 3 = 12$
$8 + 4 = 12$
$7 + 5 = 12$
$6 + 6 = 12$

$12 + 1 = 13$
$11 + 2 = 13$
$10 + 3 = 13$
$9 + 4 = 13$
$8 + 5 = 13$
$7 + 6 = 13$

$13 + 1 = 14$
$12 + 2 = 14$
$11 + 3 = 14$
$10 + 4 = 14$
$9 + 5 = 14$
$8 + 6 = 14$
$7 + 7 = 14$

$14 + 1 = 15$
$13 + 2 = 15$
$12 + 3 = 15$
$11 + 4 = 15$
$10 + 5 = 15$
$9 + 6 = 15$
$8 + 7 = 15$

$15 + 1 = 16$
$14 + 2 = 16$
$13 + 3 = 16$
$12 + 4 = 16$
$11 + 5 = 16$
$10 + 6 = 16$
$9 + 7 = 16$
$8 + 8 = 16$

$16 + 1 = 17$
$15 + 2 = 17$
$14 + 3 = 17$
$13 + 4 = 17$
$12 + 5 = 17$
$11 + 6 = 17$
$10 + 7 = 17$
$9 + 8 = 17$

$17 + 1 = 18$
$16 + 2 = 18$
$15 + 3 = 18$
$14 + 4 = 18$
$13 + 5 = 18$
$12 + 6 = 18$
$11 + 7 = 18$
$10 + 8 = 18$
$9 + 9 = 18$

$18 + 1 = 19$
$17 + 2 = 19$
$16 + 3 = 19$
$15 + 4 = 19$
$14 + 5 = 19$
$13 + 6 = 19$
$12 + 7 = 19$
$11 + 8 = 19$
$10 + 9 = 19$

Tip: Use the method shown in the following examples to mentally add a digit to a larger number.

- $24 + 8 = ?$
  $24 + 6 = 30$, so the answer is 32.
- $47 + 9 = ?$
  $47 + 3 = 50$, so the answer is 56.
- $36 + 5 = ?$
  $36 + 4 = 40$, so the answer is 41.

The result of adding numbers is called their SUM.

Adding 0 to any number doesn't change its value.

- $342 + 0 = 342$

More than two digits can be added mentally.

- $3 + 5 + 6 = 14$ First, mentally add 3 plus 5 to get 8.

  Then, mentally add 8 plus 6.

The order in which numbers are added doesn't matter.

- $3 + 6 + 5 = 14$ First add 3 plus 6 to get 9.

  Then add 9 plus 5.
- $5 + 3 + 6 = 14$ First add 5 plus 3 to get 8.

  Then add 8 plus 6.
- $6 + 5 + 3 = 14$ First add 6 plus 5 to get 11.

  Then add 11 plus 3.

Tip: Mentally add digits whose sum is 10 before adding any remaining digits.

- $2 + 6 + 8 + 3 + 4 = ?$ $2 + 8 = 10$ and $6 + 4 = 10$, so the sum is 23.

Add whole numbers column by column, starting with the ones column.

Steps

1. Line up digits having the same place value in the same column.

   To find the sum of 214, 31, and 2, first write:

$$\begin{array}{r} 214 \\ 31 \\ +\ 2 \\ \hline \end{array}$$

2. Add the digits in the ones column. Repeat with the remaining columns, moving left one column at a time.

$$\begin{array}{r} 214 \\ 31 \\ +\ 2 \\ \hline 7 \end{array}$$

Start by mentally adding the digits in the ones column. $4 + 1 + 2 = 7$. Write 7 below the line under the digits in the ones column.

$$\begin{array}{r} 214 \\ 31 \\ +\ 2 \\ \hline 47 \end{array}$$

Next, mentally add the digits in the tens column: $1 + 3 = 4$. Write 4 below the line under the digits in the tens column.

$$\begin{array}{r} 214 \\ 31 \\ +\ 2 \\ \hline 247 \end{array}$$

Complete the result by mentally adding the digits in the hundreds column.

- IF THE SUM OF A COLUMN REACHES 10 OR MORE, **CARRY** the digit in the tens place of this sum into the next larger place value.

  - *1* ←carry

    ```
     123
      45
    + 6
    ----
     174
    ```

    Mentally add the digits in the ones column: 3 + 5 + 6 = 14. Write 4 below the line under the digits in the ones column and carry the 1.

    Mentally add the digits in the tens column. The 1 carried into the tens column is added to 2 and 4 to get 7. Write 7 below the line under the digits in the tens column.

    The 1 in the hundreds column has nothing to be added to, so write a 1 in the result.

  - *12* ←carries

    ```
     547
      37
    + 49
    ----
     633
    ```

    Mentally add the digits in the ones column: 7 + 7 + 9 = 23. Write 3 below the line under the digits in the ones column and carry the 2.

    Mentally add the digits in the tens column: 2 + 4 + 3 + 4 = 13. Write the 3 below the line under the digits in the ones column and carry the 1.

    The 1 carried into the hundreds column is added to the 5 to get 6.

Subtracting Whole Numbers

Tip: In order to effectively subtract numbers, you should be able to subtract any digit from a larger one, as well as immediately respond with the answer to each of the following subtraction facts.

$11 - 2 = 9$	$12 - 3 = 9$	$13 - 4 = 9$	$14 - 5 = 9$
$11 - 3 = 8$	$12 - 4 = 8$	$13 - 5 = 8$	$14 - 6 = 8$
$11 - 4 = 7$	$12 - 5 = 7$	$13 - 6 = 7$	$14 - 7 = 7$
$11 - 5 = 6$	$12 - 6 = 6$	$13 - 7 = 6$	$14 - 8 = 6$
$11 - 6 = 5$	$12 - 7 = 5$	$13 - 8 = 5$	$14 - 9 = 5$
$11 - 7 = 4$	$12 - 8 = 4$	$13 - 9 = 4$	
$11 - 8 = 3$	$12 - 9 = 3$		
$11 - 9 = 2$			

$11 - 2 = 9$	$12 - 3 = 9$	$13 - 4 = 9$	$14 - 5 = 9$
$11 - 3 = 8$	$12 - 4 = 8$	$13 - 5 = 8$	$14 - 6 = 8$
$11 - 4 = 7$	$12 - 5 = 7$	$13 - 6 = 7$	$14 - 7 = 7$
$11 - 5 = 6$	$12 - 6 = 6$	$13 - 7 = 6$	$14 - 8 = 6$
$11 - 6 = 5$	$12 - 7 = 5$	$13 - 8 = 5$	$14 - 9 = 5$
$11 - 7 = 4$	$12 - 8 = 4$	$13 - 9 = 4$	
$11 - 8 = 3$	$12 - 9 = 3$		
$11 - 9 = 2$			

$15 - 6 = 9$	$16 - 7 = 9$	$17 - 8 = 9$	$18 - 9 = 9$
$15 - 7 = 8$	$16 - 8 = 8$	$17 - 9 = 8$	
$15 - 8 = 7$	$16 - 9 = 7$		
$15 - 9 = 6$			

The result of subtracting one number from another is called their DIFFERENCE.

Subtracting 0 from a number doesn't change its value.

- $8 - 0 = 8$

Subtracting a number from itself results in 0.

- $5 - 5 = 0$

Subtract whole numbers column by column, starting with the ones column.

Steps

1. Line up digits having the same place value in the same column, writing the larger number above the smaller one.

   To find the difference between 345 and 21, first write:

   $$\begin{array}{r} 345 \\ \underline{-21} \end{array}$$

2. Subtract the digits in the ones column. Repeat with the remaining columns, moving left one column at a time.

$\begin{array}{r} 345 \\ \underline{-21} \\ 4 \end{array}$	Start by mentally subtracting the digits in the ones column. 5 - 1 = 4. Write 4 below the line under the digits in the ones column.
$\begin{array}{r} 345 \\ \underline{-21} \\ 24 \end{array}$	Mentally subtract the digits in the tens column: 4 - 2 = 2. Write 2 below the line under the digits in the tens column.
$\begin{array}{r} 345 \\ \underline{-21} \\ 324 \end{array}$	Mentally subtract the digits in the hundreds column: 3 - 0 = 3. Write 3 below the line under the digits in the hundreds column.

- IF THE UPPER DIGIT IS LESS THAN THE DIGIT BEING SUBTRACTED, **BORROW** 10 from the next place value on the left.

```
 1 6 2      4 cannot be subtracted from 2.
-  1 4
```

```
   5
 1 6 12     Borrow a 10 from the six tens. The 6
-  1 4      becomes 5.
 1 4 8      Add the 10 that was borrowed to
            the 2, making it 12.
            Subtract 12 - 4 to get 8.
            Subtract 5 - 1 to get 4.
            Subtract 1 - 0 to get 1.
```

Check the answer to a subtraction problem by mentally adding up from the answer to see if the result is the top number.

- 345 − 21 = 324

  CHECK: 4 + 1 = 5
  2 + 2 = 4
  3 + 0 = 3
  324 + 21 = 345
  Therefore, 324 is the correct answer.

- 162 − 14 = 148

  CHECK: 8 + 4 = 12 (carry the 1)
  4 + 1 + the carry = 6
  1 + 0 = 1
  148 + 14 = 162
  Therefore, 148 is the correct answer.

Multiplying Whole Numbers

Tip: In order to effectively multiply numbers, you should be able to immediately respond with the answer to each of the following multiplication facts.

$1 \times 1 = 1$	$2 \times 1 = 2$	$3 \times 1 = 3$	$4 \times 1 = 4$	$5 \times 1 = 5$
$1 \times 2 = 2$	$2 \times 2 = 4$	$3 \times 2 = 6$	$4 \times 2 = 8$	$5 \times 2 = 10$
$1 \times 3 = 3$	$2 \times 3 = 6$	$3 \times 3 = 9$	$4 \times 3 = 12$	$5 \times 3 = 15$
$1 \times 4 = 4$	$2 \times 4 = 8$	$3 \times 4 = 12$	$4 \times 4 = 16$	$5 \times 4 = 20$
$1 \times 5 = 5$	$2 \times 5 = 10$	$3 \times 5 = 15$	$4 \times 5 = 20$	$5 \times 5 = 25$
$1 \times 6 = 6$	$2 \times 6 = 12$	$3 \times 6 = 18$	$4 \times 6 = 24$	$5 \times 6 = 30$
$1 \times 7 = 7$	$2 \times 7 = 14$	$3 \times 7 = 21$	$4 \times 7 = 28$	$5 \times 7 = 35$
$1 \times 8 = 8$	$2 \times 8 = 16$	$3 \times 8 = 24$	$4 \times 8 = 32$	$5 \times 8 = 40$
$1 \times 9 = 9$	$2 \times 9 = 18$	$3 \times 9 = 27$	$4 \times 9 = 36$	$5 \times 9 = 45$
$1 \times 10 = 10$	$2 \times 10 = 20$	$3 \times 10 = 30$	$4 \times 10 = 40$	$5 \times 10 = 50$

$6 \times 1 = 6$	$7 \times 1 = 7$	$8 \times 1 = 8$	$9 \times 1 = 9$	$10 \times 1 = 10$
$6 \times 2 = 12$	$7 \times 2 = 14$	$8 \times 2 = 16$	$9 \times 2 = 18$	$10 \times 2 = 20$
$6 \times 3 = 18$	$7 \times 3 = 21$	$8 \times 3 = 24$	$9 \times 3 = 27$	$10 \times 3 = 30$
$6 \times 4 = 24$	$7 \times 4 = 28$	$8 \times 4 = 32$	$9 \times 4 = 36$	$10 \times 4 = 40$
$6 \times 5 = 30$	$7 \times 5 = 35$	$8 \times 5 = 40$	$9 \times 5 = 45$	$10 \times 5 = 50$
$6 \times 6 = 36$	$7 \times 6 = 42$	$8 \times 6 = 48$	$9 \times 6 = 54$	$10 \times 6 = 60$
$6 \times 7 = 42$	$7 \times 7 = 49$	$8 \times 7 = 56$	$9 \times 7 = 63$	$10 \times 7 = 70$
$6 \times 8 = 48$	$7 \times 8 = 56$	$8 \times 8 = 64$	$9 \times 8 = 72$	$10 \times 8 = 80$
$6 \times 9 = 54$	$7 \times 9 = 63$	$8 \times 9 = 72$	$9 \times 9 = 81$	$10 \times 9 = 90$
$6 \times 10 = 60$	$7 \times 10 = 70$	$8 \times 10 = 80$	$9 \times 10 = 90$	$10 \times 10 = 100$

Multiplication is repeated addition.

- $3 \times 5 = 3 + 3 + 3 + 3 + 3$

  $= 5 + 5 + 5$

The result of multiplying numbers is called their PRODUCT.

There are several ways to indicate multiplication.

- 12 times 3 can be written as:

  12×3

  $12 \cdot 3$

  $12 * 3$

  $12(3)$

  $(12)(3)$

Multiplying a number by 1 doesn't change its value.

- $342 \times 1 = 342$
- $0 \times 1 = 0$

Multiplying a number by 0 results in 0.

- $342 \times 0 = 0$
- $0 \times 0 = 0$

To find the product of three numbers, first multiply two of them. Then, multiply the result by the third number.

- $5 \times 3 \times 2 = 15 \times 2 = 30$

The order in which numbers are multiplied doesn't matter.

- $5 \times 3 \times 2 = 15 \times 2 = 30$

  $5 \times 2 \times 3 = 10 \times 3 = 30$

  $3 \times 5 \times 2 = 15 \times 2 = 30$

  $3 \times 2 \times 5 = 6 \times 5 = 30$

  $2 \times 3 \times 5 = 6 \times 5 = 30$

  $2 \times 5 \times 3 = 10 \times 3 = 30$

Multiply whole numbers column by column, starting with the ones column.

Steps

1. Line up digits having the same place value in the same column, writing the number with the most digits above the number with the least number of digits.

To find the product of 123 and 4, first write:

$$\begin{array}{r} 123 \\ \times\ 4 \\ \hline \end{array}$$

2. Multiply the top number by the digit in the ones column of the bottom number.

$$\begin{array}{r} 123 \\ \times\ 4 \\ \hline 492 \end{array}$$

$4 \times 3 = 12$. Write the 2, carry the 1.

$4 \times 2 = 8$. Add the 1 that was carried to get 9.

$4 \times 1 = 4$

- IF THE BOTTOM NUMBER HAS MORE THAN ONE DIGIT, multiply the top number by each digit, beginning to write each result under the multiplying digit.
Then add the products together.

$$\begin{array}{r} 123 \\ \times 24 \\ \hline 492 \\ 246 \\ \hline 2952 \end{array}$$

In effect, two multiplication problems are solved, and their results are added together.

$$\begin{array}{r} 123 \\ \times\ 4 \\ \hline 492 \end{array} \qquad \begin{array}{r} 123 \\ \times 20 \\ \hline 2460 \end{array}$$

$$\begin{array}{r} 492 \\ +\ 2460 \\ \hline 2952 \end{array}$$

Tip: Avoid writing a row of zeros when multiplying by 0. Merely write a single 0 for each 0 in the bottom number.

- Instead of this:

$$\begin{array}{r} 112 \\ \times 201 \\ \hline 112 \\ 000 \\ 224 \\ \hline 22512 \end{array}$$

Write this:

$$\begin{array}{r} 112 \\ \times 201 \\ \hline 112 \\ 2240 \\ \hline 22512 \end{array}$$

- Instead of this:

$$\begin{array}{r} 123 \\ \times 40 \\ \hline 000 \\ 492 \\ \hline 4920 \end{array}$$

Write this:

$$\begin{array}{r} 123 \\ \times 40 \\ \hline 4920 \end{array}$$

- Instead of this:

$$\begin{array}{r} 12332 \\ \times 12000 \\ \hline 00000 \\ 00000 \\ 00000 \\ 24664 \\ 12332 \\ \hline 147984000 \end{array}$$

Write this:

$$\begin{array}{r} 12332 \\ \times 12000 \\ \hline 24664000 \\ 12332 \\ \hline 147984000 \end{array}$$

The zeros in this example are made to stick out, making the problem similar to $12332 \times 12 \times 1000$.

Dividing Whole Numbers

Tip: In order to effectively divide numbers, you should be able to immediately respond with the answer to each of the following division facts.

$12 \div 2 = 6$	$12 \div 3 = 4$	$12 \div 4 = 3$	$12 \div 6 = 2$
$14 \div 2 = 7$	$14 \div 7 = 2$		
$15 \div 3 = 5$	$15 \div 5 = 3$		
$16 \div 2 = 8$	$16 \div 4 = 4$	$16 \div 8 = 2$	
$18 \div 2 = 9$	$18 \div 3 = 6$	$18 \div 6 = 3$	$18 \div 9 = 2$
$20 \div 4 = 5$	$20 \div 5 = 4$		
$21 \div 3 = 7$	$21 \div 7 = 3$		
$24 \div 3 = 8$	$24 \div 4 = 6$	$24 \div 6 = 4$	$24 \div 8 = 3$
$25 \div 5 = 5$			
$27 \div 3 = 9$	$27 \div 9 = 3$		
$28 \div 4 = 7$	$28 \div 7 = 4$		
$30 \div 6 = 5$	$30 \div 5 = 6$		
$32 \div 4 = 8$	$32 \div 8 = 4$		
$35 \div 5 = 7$	$35 \div 7 = 5$		
$36 \div 4 = 9$	$36 \div 6 = 6$	$36 \div 9 = 4$	
$40 \div 5 = 8$	$40 \div 8 = 5$		
$42 \div 6 = 7$	$42 \div 7 = 6$		
$45 \div 5 = 9$	$45 \div 9 = 5$		
$48 \div 6 = 8$	$48 \div 8 = 6$		
$49 \div 7 = 7$			
$54 \div 6 = 9$	$54 \div 9 = 6$		
$56 \div 7 = 8$	$56 \div 8 = 7$		
$63 \div 7 = 9$	$63 \div 9 = 7$		
$64 \div 8 = 8$			
$72 \div 8 = 9$	$72 \div 9 = 8$		
$81 \div 9 = 9$			

The result of dividing one number by another is called their QUOTIENT.

The number being divided is called the DIVIDEND.

The number being divided *into* the dividend is called the DIVISOR.

- In the expression $12 \div 3 = 4$ or $3\overline{)12}^{\,4}$:

  12 is the dividend.

  3 is the divisor.

  4 is the quotient.

There are several ways to indicate division.

- 12 divided by 3 can be written as:

  $12 \div 3$

  $\frac{12}{3}$

  12/3

  $3\overline{)12}$

Dividing any number by 1 doesn't change its value.

- $342 \div 1 = 342$

Dividing 0 by any number other than 0 results in zero.

- $0 \div 5 = 0$

DIVIDING BY ZERO is not permitted.

We know that $\frac{6}{3} = 2$ only because $6 = 3 \times 2$.

If $\frac{6}{0}$ = [some number n] then 6 would equal $0 \times n$.
But 0 times any number is 0, so 6 would have to equal 0.

If $\frac{0}{0}$ = [some number n] then 0 would equal $0 \times n$.
But 0 times any number is 0, so n could be any number.

This confusion is avoided by saying that there is no answer when dividing by zero.

- $\frac{6}{0}$ has **No Answer.**
- $\frac{0}{0}$ has **No Answer.**

Dividing whole numbers could leave a REMAINDER.

- $6 \div 3 = 2$ with no remainder.
- $7 \div 2 = 3$ with a remainder of 1. $7 \div 2 = 3r1$
- $14 \div 5 = 2$ with a remainder of 4. $14 \div 5 = 2r4$

Check the answer to a division problem by multiplying the quotient by the divisor and then adding any remainder. The result should equal the dividend.

- $3\overline{)6}$ with quotient 2 — CHECK: $3 \times 2 = 6$
- $2\overline{)7}$ with quotient 3 r1 — CHECK: $2 \times 3 = 6$, $6 + 1 = 7$
- $14 \div 5 = 2r4$ — CHECK: $5 \times 2 = 10$, $10 + 4 = 14$

There are two ways to divide one number by another: short division and long division.

Use SHORT DIVISION to divide any whole number by a single digit.

- $2\overline{)174}$ with answer 87

  2 does not divide 1, so 2 must be divided into 17. The result of 8 is written in the answer.

  The remainder of 1 is **carried**, and written next to the 4 ($^{1}4$).

  2 is divided into 14 giving 7, which is written in the answer.

  CHECK:
  $$\begin{array}{r} 87 \\ \underline{\times 2} \\ 174 \end{array}$$

-

- $2\overline{)165}$ with answer 82 r1

  2 does not divide 1, so 2 must be divided into 16. The result of 8 is written in the answer.

  When 2 is divided into 5 the result is 2, which is written in the answer.

  The remainder of 1 is written as r1 in the answer.

  CHECK:
  $$\begin{array}{r} 82 \\ \underline{\times 2} \\ 164 \\ \underline{+1} \\ 165 \end{array}$$

Use LONG DIVISION to divide any whole number by a number that has two or more digits.

- $$\begin{array}{r} 54\ r21 \\ 25\overline{)1371} \\ -\underline{125\downarrow} \\ 121 \\ -\underline{100} \\ 21 \end{array}$$

  25 does not divide 1 or 13, so 25 must be divided into 137. Since five quarters equal $1.25, 5 is written above the line over the 7.
  25 is multiplied by 5 and 125 is written beneath 137.
  137 - 125 = 12
  The next digit (1) is brought down and written next to the 12.
  25 divides 121 four times, so 4 is written in the answer.
  $4 \times 25 = 100$.
  Subtracting 100 from 121 leaves a remainder of 21.

  CHECK:
  $$\begin{array}{r} 54 \\ \underline{\times\ 25} \\ 270 \\ \underline{108} \\ 1350 \\ \underline{+\ 21} \\ 1371 \end{array}$$

If it is hard to see how many times a large divisor can go into a number, mentally round the numbers.

- The problem $37\overline{)132}$ can be estimated by rounding the numbers: $40\overline{)130} \approx 3$. (≈ means "APPROXIMATELY".)

  $$\begin{array}{r} 3 \\ 37\overline{)132} \end{array}$$

  Begin by placing the estimate in the quotient. Then finish dividing.

  Tip: If the estimate is too big or too small, increase or decrease it by 1.

Divisibility Rules

A number is divisible by 2 if it ends in 0, 2, 4, 6, or 8. A number divisible by 2 is an EVEN NUMBER. A number that is not divisible by 2 is an ODD NUMBER.

A number is divisible by 10 if its last digit is 0.

- 720, 40, and 3450 are divisible by 10.

A number is divisible by 5 if its last digit is 0 or 5.

- 365, 130, and 75 are divisible by 5.

A number is divisible by 4 if its last two digits are divisible by 4.

- 3080, 716, and 2308 are divisible by 4.

A number is divisible by 8 if its last three digits are divisible by 8.

- 1,800, 2,016, and 1,000,000 are divisible by 8.

A number is divisible by 3 if the sum of its digits is divisible by 3.

- 123, 2811, and 1218 are divisible by 3.

A number is divisible by 9 if the sum of its digits is divisible by 9.

- 252, 36, and 1089 are divisible by 9.

Multiples of a Whole Number

A MULTIPLE of a whole number is any number that it divides exactly.

- Multiples of 3 include 3, 6, 9, 12, and 15.

  $3 \cdot 1 = 3$

  $3 \cdot 2 = 6$

  $3 \cdot 3 = 9$

  $3 \cdot 4 = 12$

  $3 \cdot 5 = 15$

- Of the numbers 9, 700, 71, 14, 57, only 700 and 14 are multiples of 7.

  Dividing 9, 71, or 57 by 7 leaves a remainder.

- Some multiples of 20 are:

  20, 40, 60, 80, 100, 120, and 400.

When a number is a multiple of several whole numbers, it is called a COMMON MULTIPLE of the numbers.

- 24 is a common multiple of
  1, 2, 3, 4, 6, 8, 12, and 24

  $1 \times 24 = 24$

  $2 \times 12 = 24$

  $3 \times 8 = 24$

  $4 \times 6 = 24$

- 15 is a common multiple of 1, 3, 5, and 15
- 100 is a common multiple of
  1, 2, 4, 5, 10, 20, 25, 50, and 100
- To find common multiples of 4 and 6, write multiples of the greater number, 6.
  6, 12, 18, 24, 30, 36, 42, ...

  Select those multiples of 6 that are also multiples of 4 (the lesser number): 12, 24, and 36.

  Therefore, 12, 24, and 36 are common multiples of both 4 and 6.

The LEAST COMMON MULTIPLE of several whole numbers is called their LCM.

- To find the least common multiple of 8 and 12, write a few multiples of the greater number, 12.
  12, 24, 36, 48, ...

  Select the smallest multiple of 12 that is also a multiple of 8.

  The least common multiple (LCM) of 8 and 12 is 24.

Every group of whole numbers has an LCM.

MULTIPLES			
	36	36	Common multiple
	30		
		27	
	24		
	18	18	Least Common Multiple
	12		
		9	
	6		
	6	9	

Factors of a Whole Number

A FACTOR of a whole number is a number that exactly divides it. Every whole number has a limited number of factors.

- To find all factors of 24, find every pair of numbers whose product is 24.

$$\begin{aligned} 24 &= 1 \times 24 \\ &= 2 \times 12 \\ &= 3 \times 8 \\ &= 4 \times 6 \end{aligned}$$

- To find all factors of 12, find every pair of numbers whose product is 12.

$$\begin{aligned} 12 &= 1 \cdot 12 \\ &= 2 \cdot 6 \\ &= 3 \cdot 4 \end{aligned}$$

- $$\begin{aligned} 20 &= 1 \times 20 \\ &= 2 \times 10 \\ &= 4 \times 5 \end{aligned}$$

- $$\begin{aligned} 9 &= 1 \times 9 \\ &= 3 \times 3 \end{aligned}$$

  9 has three factors: 1, 3, and 9

When a number is a factor of several whole numbers it is called a COMMON FACTOR.

- 1, 2, 4, and 8 exactly divide both 40 and 64. Therefore, 1, 2, 4, and 8 are common factors of 40 and 64.

The greatest common factor of several whole numbers is called their GCF.

- To find the greatest common factor of 9, 24, and 36, write factors of the smallest number, 9: 1, 3, and 9.

  Select those factors of 9 that are also factors of 24 and 36: 1 and 3.

  Since 1 and 3 are common factors of all three numbers, the greatest common factor (GCF) of 9, 24, and 36 is 3.

- The GCF of 2 and 15 is 1.

Every group of whole numbers has a GCF.

FACTORS	6	9	
		9	
	6		
	3	3	Greatest Common Factor
	2		
	1	1	Common factor

Prime and Composite Numbers

Any whole number that has only two factors, 1 and itself, is a PRIME NUMBER.

- The prime numbers less than 100 are:
  2, 3, 5, 7, 11, 13, 17, 19, 23, 29, 31, 37, 41,
  43, 47, 53, 59, 61, 67, 71, 73, 79, 83, 89, and 97

0 and 1 are not prime numbers.

2 is the lowest prime number and is the only *even* prime number.

When 1 is the only common factor of several numbers, they are said to be RELATIVELY PRIME.

- 2 and 15 are relatively prime.
- 3, 8, and 25 are relatively prime.

A COMPOSITE NUMBER is a whole number greater than 2 that is not prime.

- Composite numbers include:

  4 (factors: 1, 2, and 4)
  10 (factors: 1, 2, 5, and 10)
  12 (factors: 1, 2, 3, 4, 6, and 12)
  22 (factors: 1, 2, 11 and 22)

Every composite number is the product of prime numbers.

- $4 = 2 \times 2$
- $6 = 2 \times 3$
- $8 = 2 \times 2 \times 2$
- $10 = 2 \times 5$
- $12 = 2 \times 2 \times 3$
- $22 = 2 \times 11$

Use a PRIME FACTOR TREE to find the prime factors of a composite number.

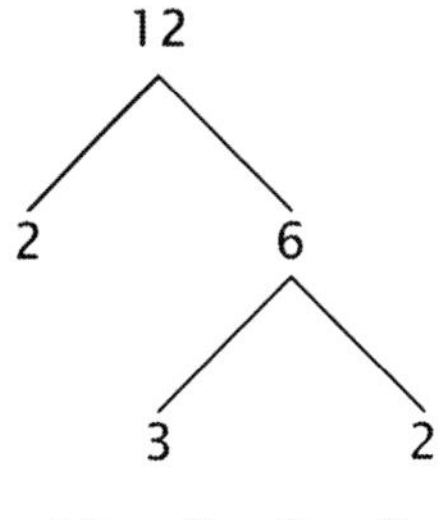

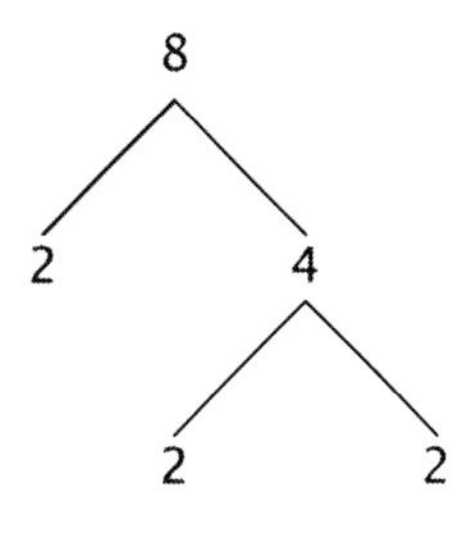

$12 = 2 \times 3 \times 2$

$8 = 2 \times 2 \times 2$

NOTE: IF 12 WAS FIRST FACTORED AS 3×4, AND THEN 4 FACTORED AS 2×2, THE PRIME FACTORS WOULD BE THE SAME.

Prime factor trees can be used to find both the greatest common factor and least common multiple of several numbers.

- The product of primes common to each number is the GCF of 12 and 8.

  $\mathbf{2 \times 2}$

  4 is the greatest common factor (GCF) of 12 and 8.

- The smallest product of primes that includes all the prime factors of each number is the LCM of 12 and 8.

  $\mathbf{3 \times 2 \times 2 \times 2}$

  24 is the least common multiple (LCM) of 12 and 8.

Fractions

Fraction. Numerator. Denominator. Common denominator. Proper fraction. Improper fraction. Mixed number. Equal fractions. Cross-products. Simplest form. Reduced. Lowest terms. Higher terms. Least common denominator (LCD). Canceling. Reciprocal. Complex fraction.

Fraction, Numerator, and Denominator

A fraction represents one or more equal parts of a whole.

- $\frac{3}{4}$ is a fraction that represents three parts of a whole that was divided into four equal parts. It is read as "three fourths" or "three quarters".

The number above the line in a fraction is called the NUMERATOR of the fraction. The number below the line in a fraction is called the DENOMINATOR of the fraction.

- The fraction $\frac{3}{4}$ has a numerator of 3 and denominator of 4.

When the numerator and denominator of a fraction are equal, the fraction equals 1.

- $\frac{2}{2}$ (two halves) $= 1$
- $\frac{10}{10}$ (ten tenths) $= 1$

There are as many ways to write 1 as a fraction as there are numbers.

- $1 = \frac{1}{1} = \frac{2}{2} = \frac{3}{3}$, etc.

Fractions that have the same denominator are said to have a COMMON DENOMINATOR.

- $\frac{3}{5}$ and $\frac{1}{5}$ have a common denominator.

Fractions as Division

A fraction can be treated as a division statement.

Obvious examples include:

- $\frac{1}{2} = 1 \div 2 = 2\overline{)1}$
- $\frac{1}{3} = 1 \div 3 = 3\overline{)1}$
- $\frac{1}{4} = 1 \div 4 = 4\overline{)1}$

Less obvious are these examples:

- $\frac{2}{3} = 2 \div 3 = 3\overline{)2}$

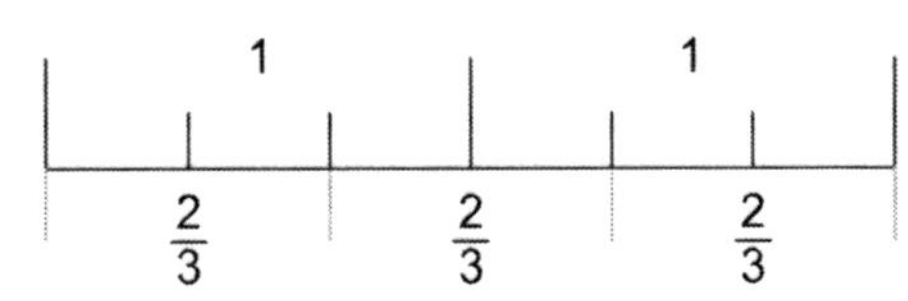

- $\frac{3}{4} = 3 \div 4 = 4\overline{)3}$

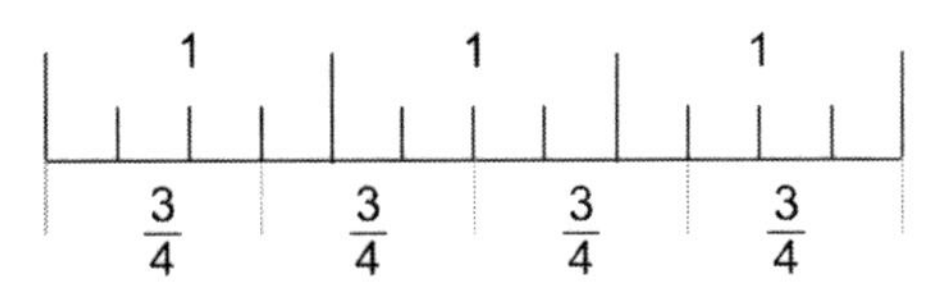

Any REMAINDER to a division problem should be written as a fraction having a denominator equal to the divisor.

- Instead of this: $2\overline{)165}$ = 82 r1

  Write this: $2\overline{)165}$ = $82\frac{1}{2}$

  The $\frac{1}{2}$ in the quotient results from dividing the remainder by 2.

Proper & Improper Fractions, Mixed Numbers

When a fraction has a numerator that is less than the denominator, its value is less than 1 and it is called a PROPER FRACTION.

- $\frac{4}{5}$ and $\frac{1}{8}$ are proper fractions.

When a fraction has a numerator that is *not* less than the denominator, its value is greater than or equal to 1 and it is called an IMPROPER FRACTION.

- $\frac{10}{2}$, $\frac{6}{5}$ and $\frac{8}{8}$ are improper fractions.

A MIXED NUMBER is a number that has a whole number in addition to a proper fraction.

- $5\frac{3}{4}$ is a mixed number that represents 5 whole objects and $\frac{3}{4}$ of another object. It is read as "five and three quarters".

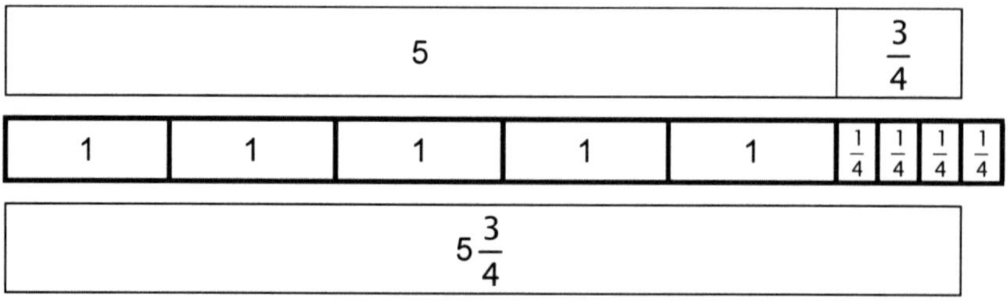

5					$\frac{3}{4}$			
1	1	1	1	1	$\frac{1}{4}$	$\frac{1}{4}$	$\frac{1}{4}$	$\frac{1}{4}$
$5\frac{3}{4}$								

Fractional Equivalence and Conversions

Equal fractions

When two different fractions represent the same quantity, they are equivalent, or EQUAL FRACTIONS.

- $\frac{3}{4} = \frac{6}{8}$

1							
$\frac{1}{4}$		$\frac{1}{4}$		$\frac{1}{4}$		$\frac{1}{4}$	
$\frac{1}{8}$	$\frac{1}{8}$	$\frac{1}{8}$	$\frac{1}{8}$	$\frac{1}{8}$	$\frac{1}{8}$	$\frac{1}{8}$	$\frac{1}{8}$

- $\frac{3}{6} = \frac{1}{2}$

1					
$\frac{1}{6}$	$\frac{1}{6}$	$\frac{1}{6}$	$\frac{1}{6}$	$\frac{1}{6}$	$\frac{1}{6}$
$\frac{1}{2}$			$\frac{1}{2}$		

Multiplying or dividing both numerator and denominator of a fraction by the same number results in an equivalent fraction.

- $\frac{3}{4} = \frac{3 \times 2}{4 \times 2} = \frac{6}{8}$
- $\frac{3}{6} = \frac{3 \div 3}{6 \div 3} = \frac{1}{2}$

When two fractions are equal, their CROSS-PRODUCTS are equal.

- $\frac{3}{4} = \frac{6}{8}$ $6 \times 4 = 8 \times 3$
- $\frac{1}{2} = \frac{12}{24}$ $1 \times 24 = 2 \times 12$

Writing a fraction in lowest terms

A fraction is in SIMPLEST FORM (or REDUCED TO LOWEST TERMS) when 1 is the only common factor of its numerator and denominator.

- $\frac{1}{8}$, $\frac{3}{8}$, $\frac{5}{8}$, and $\frac{7}{8}$ are fractions in lowest terms.
- $\frac{2}{8}$, $\frac{4}{8}$, $\frac{6}{8}$, and $\frac{8}{8}$ are fractions that are not in lowest terms.

To reduce a fraction to lowest terms, divide its numerator and denominator by their greatest common factor (GCF).

- $\frac{2}{8} = \frac{1}{4}$ The 2 and the 8 are both divided by 2.
- $\frac{4}{8} = \frac{1}{2}$ The 4 and the 8 are both divided by 4.

Tip: If you don't notice the GCF, divide by any factor that you do notice. Then continue to simplify the fraction.

$$\frac{4}{8} = \frac{2}{4} = \frac{1}{2}$$

Writing a fraction in higher terms

Any fraction can be written as another fraction that has a greater numerator and denominator.

NOTE: YOU NEED TO KNOW THE DENOMINATOR OF THE NEW FRACTION BEFORE YOU BEGIN. IT SHOULD BE A MULTIPLE OF THE DENOMINATOR IN THE INITIAL FRACTION.

Steps

1. Divide the new denominator by the denominator of the original fraction.
2. Multiply this number by the numerator of the original fraction and write the result as the numerator of the new fraction.

- $\frac{1}{8} = \frac{\quad}{24}$ 8 into 24 is 3; $1 \times 3 = 3$; $\frac{1}{8} = \frac{3}{24}$
- $\frac{2}{3} = \frac{\quad}{15}$ 3 into 15 is 5; $2 \times 5 = 10$; $\frac{2}{3} = \frac{10}{15}$
- $\frac{3}{10} = \frac{\quad}{40}$ 10 into 40 is 4; $3 \times 4 = 12$; $\frac{3}{10} = \frac{12}{40}$

Equivalent improper fractions, mixed numbers & whole numbers

Every improper fraction is equal to either a whole number or mixed number.

1				1				1			
$\frac{1}{4}$	$\frac{1}{4}$	$\frac{1}{4}$	$\frac{1}{4}$	$\frac{1}{4}$	$\frac{1}{4}$	$\frac{1}{4}$	$\frac{1}{4}$	$\frac{1}{4}$	$\frac{1}{4}$	$\frac{1}{4}$	$\frac{1}{4}$

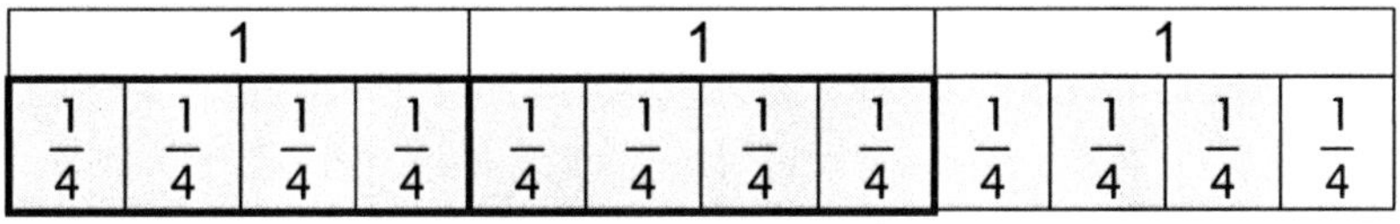

$\frac{8}{4} = 2$ $\quad$ $\frac{11}{4} = 2\frac{3}{4}$

Every mixed number is equal to an improper fraction.

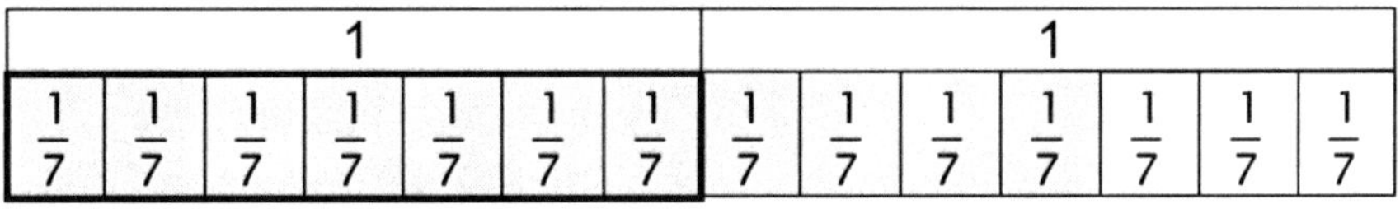

1							1						
$\frac{1}{7}$	$\frac{1}{7}$	$\frac{1}{7}$	$\frac{1}{7}$	$\frac{1}{7}$	$\frac{1}{7}$	$\frac{1}{7}$	$\frac{1}{7}$	$\frac{1}{7}$	$\frac{1}{7}$	$\frac{1}{7}$	$\frac{1}{7}$	$\frac{1}{7}$	$\frac{1}{7}$

$1\frac{4}{7} = \frac{11}{7}$

Writing an improper fraction as an equivalent whole number or mixed number

Divide the denominator into the numerator. Write the resulting whole number. Any remainder becomes the numerator of a fraction in a mixed number.

- $\frac{6}{2} = 3$ — 2 into 6 is 3, with no remainder.
- $\frac{11}{7} = 1\frac{4}{7}$ — 7 into 11 is 1, with remainder 4.
- $\frac{27}{4} = 6\frac{3}{4}$ — 4 into 27 is 6, with remainder 3.

An improper fraction is written in SIMPLEST FORM when it is converted to an equivalent mixed number or whole number.

Writing a whole number as an equivalent improper fraction

To write a whole number as an improper fraction, either write a 1 as its denominator, or write it in higher terms.

- $8 = \frac{8}{1}$
- $5 = \frac{5}{1}$
- $12 = \frac{12}{1}$

NOTE:

THESE IMPROPER FRACTIONS ARE READ AS "8 OVER 1", "5 OVER 1", AND "12 OVER 1".

THEY ARE CORRECT EXPRESSIONS OF DIVISION BECAUSE ANY NUMBER DIVIDED BY ONE REMAINS UNCHANGED.

- $1 = \frac{1}{1} = \frac{?}{5} = \frac{5}{5}$
- $8 = \frac{8}{1} = \frac{?}{2} = \frac{16}{2}$
- $8 = \frac{8}{1} = \frac{?}{3} = \frac{24}{3}$

Writing a mixed number as an equivalent improper fraction

Since a mixed number consists of a whole number and a fraction, convert the whole number into a fraction and then add it to the existing one.

Steps

1. Make the denominator of the improper fraction the same as that in the fraction of the mixed number.
2. To calculate the numerator of the improper fraction:
   (1) Multiply the denominator of the fraction by the whole number to determine the number of parts in the wholes.
   (2) Add this number to the numerator of the fraction in the mixed number.

- $1\frac{4}{7} = \frac{?}{7}$ 1 times 7 is 7 ($1 = \frac{7}{7}$)
  $= \frac{11}{7}$ 7 plus 4 is 11.

- $6\frac{3}{4} = \frac{?}{4}$ 6 times 4 is 24 ($6 = \frac{24}{4}$)
  $= \frac{27}{4}$ 24 plus 3 is 27.

Replacing several fractions with equal fractions having their least common denominator

The LEAST COMMON DENOMINATOR (LCD) of several fractions is found by calculating the least common multiple (LCM) of their denominators. The LCD is then used as the denominator of equal fractions written in higher terms.

- To find the least common denominator of $\frac{1}{2}$, $\frac{5}{6}$, and $\frac{3}{8}$, find the least common multiple of 2, 6, and 8, which is 24. (Prime factor trees yield $2 \times 2 \times 2 \times 3 = 24$.)

  Then, to find fractions equivalent to $\frac{1}{2}$, $\frac{5}{6}$, and $\frac{3}{8}$ that have a common denominator of 24, write the fractions in higher terms:

  $\frac{1}{2} = \frac{}{24}$ (Since $2 \times 12 = 24$, multiply 1×12.) $\frac{1}{2} = \frac{12}{24}$

  $\frac{5}{6} = \frac{}{24}$ (Since $6 \times 4 = 24$, multiply 5×4.) $\frac{5}{6} = \frac{20}{24}$

  $\frac{3}{8} = \frac{}{24}$ (Since $8 \times 3 = 24$, multiply 3×3.) $\frac{3}{8} = \frac{9}{24}$

- To find the least common denominator of $\frac{3}{4}$ and $\frac{1}{10}$, find the least common multiple of 4 and 10, which is 20. (Prime factor trees yield $2 \times 2 \times 5 = 20$.)

  Then, to find fractions equivalent to $\frac{3}{4}$ and $\frac{1}{10}$ that have a common denominator of 20, write the fractions in higher terms:

  $\frac{3}{4} = \frac{}{20}$ (Since $4 \times 5 = 20$, multiply 3×5.) $\frac{3}{4} = \frac{15}{20}$

  $\frac{1}{10} = \frac{}{20}$ (Since $10 \times 2 = 20$, multiply 1×2.) $\frac{1}{10} = \frac{2}{20}$

Ordering Fractions

To order fractions that have a common denominator, write them in the order of their numerators.

- The fractions $\frac{1}{7}, \frac{3}{7}, \frac{4}{7}, \frac{6}{7}$ are in ascending order.
- The fractions $\frac{4}{5}, \frac{2}{5}, \frac{1}{5}$ are in descending order.

To order fractions that have different denominators, first replace them with equivalent fractions that have a common denominator. Then, order the fractions that have a common denominator.

Steps

1. Find the least common denominator of the fractions.

   To order $\frac{5}{9}, \frac{2}{5}, \frac{2}{3}$, first find the least common denominator of the fractions, which is 45.

   (Prime factor trees yield $3 \times 3 \times 5 = 45$.)

2. Replace the fractions with equivalent fractions that have the common denominator.

   $$\frac{5}{9} = \frac{25}{45}, \quad \frac{2}{5} = \frac{18}{45}, \quad \frac{2}{3} = \frac{30}{45}$$

3. Write the equivalent fractions in the order of their numerators.

   $$\frac{18}{45}, \frac{25}{45}, \frac{30}{45}$$

4. Write the original fractions in order.

   $$\frac{2}{5}, \frac{5}{9}, \frac{2}{3}$$

Adding Numbers That Include Fractions

Adding fractions that have a common denominator

Fractions can be added only when they have a common denominator.

Steps

1. Begin by writing the fractions beneath one another.
2. Add the numerators of the fractions and use the sum as the numerator of the result.
3. Use the denominator of the fractions being added as the denominator of the result.

- $$\begin{array}{r} \frac{1}{5} \\ +\ \frac{2}{5} \\ \hline \frac{3}{5} \end{array}$$

- $$\begin{array}{r} \frac{10}{17} \\ +\ \frac{3}{17} \\ \hline \frac{13}{17} \end{array}$$

Adding numbers that include fractions, whole numbers, and mixed numbers

Adding numbers that include fractions, whole numbers, and mixed numbers is similar to adding fractions. The only difference is that the whole numbers are lined up in another column and added separately.

Steps

1. Begin by writing the numbers beneath one another.
2. Add the fractions.
3. Add the whole numbers.

- $$\begin{array}{r} 5\frac{3}{7} \\ 2 \\ \frac{1}{7} \\ +\ 3\frac{2}{7} \\ \hline 10\frac{6}{7} \end{array}$$

Adding fractions that do not have a common denominator

When fractions have different denominators, replace them with equivalent fractions that have the least common denominator. Then add as usual.

Steps

1. Begin by writing the numbers beneath one another.
2. Find the least common denominator.
3. Write the fractions in higher terms, using the least common denominator in the equivalent fractions.
4. Add the numbers whose fractions now have a common denominator.

- $$\begin{array}{rcl} \frac{4}{9} & = & \frac{16}{36} \\ \frac{1}{4} & = & \frac{9}{36} \\ +\frac{1}{6} & = & \frac{6}{36} \\ \hline & & \frac{31}{36} \end{array}$$

- $$\begin{array}{rcl} 2\frac{1}{3} & = & 2\frac{5}{15} \\ +5\frac{2}{5} & = & 5\frac{6}{15} \\ \hline & & 7\frac{11}{15} \end{array}$$

Writing the answer to an addition problem in simplest terms

After adding fractions, if the result contains a fraction that is not in lowest terms, reduce the fraction to lowest terms.

- $$\begin{array}{r} \frac{1}{5} \\ +\ \frac{2}{5} \\ \hline \frac{3}{5} \end{array}$$

  NOTE: THIS ANSWER IS IN SIMPLEST FORM AND CANNOT BE REDUCED.
  IT IS LEFT ALONE.

- $$\begin{array}{r} \frac{4}{15} \\ +\ \frac{8}{15} \\ \hline \frac{12}{15} = \frac{4}{5} \end{array}$$

- $$\begin{array}{r} 2\frac{3}{8} \\ 2\frac{1}{8} \\ 1\frac{1}{8} \\ +\ \frac{1}{8} \\ \hline 5\frac{6}{8} = 5\frac{3}{4} \end{array}$$

After adding fractions, if the result contains an improper fraction, replace it with an equivalent whole number or mixed number.

- $$\begin{array}{r} \frac{2}{7} \\ + \frac{5}{7} \\ \hline \frac{7}{7} = 1 \end{array}$$

- $$\begin{array}{rcl} \frac{1}{3} & = & \frac{6}{18} \\ \frac{5}{6} & = & \frac{15}{18} \\ + \frac{8}{9} & = & \frac{16}{18} \\ \hline & & \frac{37}{18} = 2\frac{1}{18} \end{array}$$

▸ IF THE RESULT ALREADY INCLUDES A WHOLE NUMBER, add it to the whole number or mixed number that replaced the improper fraction.

- $$\begin{array}{rcl} 2\frac{1}{6} & = & 2\frac{5}{30} \\ 2\frac{2}{5} & = & 2\frac{12}{30} \\ + \frac{13}{30} & = & \frac{13}{30} \\ \hline & & 4\frac{30}{30} = 5 \end{array}$$

  $$(4\frac{30}{30} = 4 + \frac{30}{30} = 4 + 1 = 5)$$

- $$\begin{array}{r} 1\frac{3}{5} \\ + \frac{3}{5} \\ \hline 1\frac{6}{5} = 2\frac{1}{5} \end{array}$$

  $$(1\frac{6}{5} = 1 + \frac{6}{5} = 1 + 1\frac{1}{5} = 2\frac{1}{5})$$

After adding fractions, if the result contains an improper fraction that is not in lowest terms, change it to a mixed number and reduce the fraction.

Tip: You can change the improper fraction to a mixed number first, or reduce the fraction first.
The result will be the same.

- $$\begin{array}{rcl} \frac{1}{3} & = & \frac{6}{18} \\ \frac{8}{9} & = & \frac{16}{18} \\ +\frac{4}{6} & = & \frac{12}{18} \\ \hline & & \frac{34}{18} \end{array}$$

$\frac{34}{18}$ can be changed to a mixed number first, then reduced to lowest terms.

$$\frac{34}{18} = 1\frac{16}{18} = 1\frac{8}{9}$$

$\frac{34}{18}$ can be reduced to lowest terms first, then changed to a mixed number.

$$\frac{34}{18} = \frac{17}{9} = 1\frac{8}{9}$$

Subtracting Numbers That Include Fractions

Subtracting fractions that have a common denominator

Fractions can be subtracted only when they have a common denominator.

Steps

1. Begin by writing the smaller fraction beneath the larger one.
2. Subtract the numerator of the bottom fraction from the numerator of the top fraction. Use this answer as the numerator of the result.
3. Use the denominator of the fractions being subtracted as the denominator of the result.

- $$\begin{array}{r} \frac{5}{7} \\ -\frac{3}{7} \\ \hline \frac{2}{7} \end{array}$$

- $$\begin{array}{r} \frac{15}{99} \\ -\frac{5}{99} \\ \hline \frac{10}{99} \end{array}$$

Subtracting numbers that include fractions, whole numbers, and mixed numbers

Subtracting numbers that include fractions, whole numbers, and mixed numbers is similar to subtracting fractions. The only difference is that the whole numbers are lined up in another column and subtracted separately.

Steps

1. Begin by writing the smaller number beneath the larger one.
2. Subtract the fractions.
3. Subtract the whole numbers.

- $$\begin{array}{r} 4\frac{3}{5} \\ -\ 1\frac{1}{5} \\ \hline 3\frac{2}{5} \end{array}$$

- $$\begin{array}{r} 8\frac{5}{7} \\ -\ \frac{3}{7} \\ \hline 8\frac{2}{7} \end{array}$$

- $$\begin{array}{r} 5\frac{1}{2} \\ -\ 2\phantom{\frac{1}{2}} \\ \hline 3\frac{1}{2} \end{array}$$

Subtracting fractions that do not have a common denominator

When fractions have different denominators, replace them with equivalent fractions that have a common denominator. Then subtract as usual.

Steps

1. Begin by writing the smaller number beneath the larger one.
2. Find the least common denominator.
3. Write the fractions in higher terms, using the least common denominator in the equivalent fractions.
4. Subtract the numbers whose fractions now have a common denominator.

- $$\begin{array}{rcr} \frac{3}{8} & = & \frac{3}{8} \\ - \frac{1}{4} & = & \frac{2}{8} \\ \hline & & \frac{1}{8} \end{array}$$

- $$\begin{array}{rcr} 6\frac{4}{5} & = & 6\frac{8}{10} \\ - 2\frac{1}{2} & = & 2\frac{5}{10} \\ \hline & & 4\frac{3}{10} \end{array}$$

Subtracting fractions when the fraction to be subtracted is greater than the fraction above it

If there is no fraction in the top number, only a whole number, borrow 1 from it. Write the 1 as a fraction with numerator and denominator equal to the denominator of the fraction being subtracted.

NOTE: WHEN YOU BORROW 1, THE VALUE OF THE WHOLE NUMBER ON TOP IS DECREASED.

- $$\begin{array}{rcr} 2 & = & 1\frac{2}{2} \\ -\frac{1}{2} & = & \frac{1}{2} \\ \hline & & 1\frac{1}{2} \end{array}$$

  Because there is no fraction in the top number, 1 is borrowed from the 2.

  The borrowed 1 is written as $\frac{2}{2}$ because the fraction being subtracted has a denominator of 2.

- $$\begin{array}{rcr} 5 & = & 4\frac{7}{7} \\ -1\frac{1}{7} & = & 1\frac{1}{7} \\ \hline & & 3\frac{6}{7} \end{array}$$

  Because there is no fraction in the top number, 1 is borrowed from the 5.

  The borrowed 1 is written as $\frac{7}{7}$ because the fraction being subtracted has a denominator of 7.

- $$\begin{array}{rcr} 8 & = & 7\frac{5}{5} \\ -7\frac{3}{5} & = & 7\frac{3}{5} \\ \hline & & \frac{2}{5} \end{array}$$

  Because there is no fraction in the top number, 1 is borrowed from the 8.

  The borrowed 1 is written as $\frac{5}{5}$ because the fraction being subtracted has a denominator of 5.

If the top number has a fraction, that fraction must be added to the 1 you borrowed.

- $$\begin{array}{rcl} 8\frac{1}{4} & = & 7\frac{5}{4} \\ -\ \frac{3}{4} & = & \frac{3}{4} \\ \hline & & 7\frac{2}{4} = 7\frac{1}{2} \end{array}$$

Because $\frac{1}{4}$ is less than $\frac{3}{4}$, 1 is borrowed from the 8. The borrowed 1 is treated as $\frac{4}{4}$ since the fractions are $\frac{1}{4}$ and $\frac{3}{4}$.

Then, $\frac{4}{4}$ is mentally added to $\frac{1}{4}$, and $\frac{5}{4}$ is written as the upper fraction.

- $$\begin{array}{rcrcr} 5\frac{1}{6} & = & 5\frac{2}{12} & = & 4\frac{14}{12} \\ -\ 2\frac{3}{4} & = & 2\frac{9}{12} & = & 2\frac{9}{12} \\ \hline & & & & 2\frac{5}{12} \end{array}$$

NOTE: BORROW ONLY AFTER THE FRACTIONS HAVE A COMMON DENOMINATOR.

Because $\frac{2}{12}$ is less than $\frac{9}{12}$, 1 is borrowed from the 5 as $\frac{12}{12}$ since the fractions are $\frac{2}{12}$ and $\frac{9}{12}$. Then, $\frac{12}{12}$ is mentally added to $\frac{2}{12}$, and $\frac{14}{12}$ is written as the upper fraction.

Writing the answer to a subtraction problem in simplest terms

Tip: The result of subtracting fractions can never include an improper fraction because the top fraction can never be 1 more than the bottom fraction.

After subtracting fractions, if the result contains a fraction that is not in lowest terms, reduce it to lowest terms.

- $$\begin{array}{r} 7\frac{2}{3} \\ -\ 2\frac{1}{3} \\ \hline 5\frac{1}{3} \end{array}$$

  NOTE: THIS ANSWER IS IN SIMPLEST FORM AND CANNOT BE REDUCED. IT IS LEFT ALONE.

- $$\begin{array}{r} 3\frac{3}{4} \\ -\ 1\frac{1}{4} \\ \hline 2\frac{2}{4} \end{array} = 2\frac{1}{2}$$

- $$\begin{array}{rcl} \frac{8}{9} & = & \frac{16}{18} \\ -\ \frac{4}{6} & = & \frac{12}{18} \\ \hline & & \frac{4}{18} = \frac{2}{9} \end{array}$$

Multiplying Numbers That Include Fractions

Multiplying fractions

When multiplying fractions, write them in a single row, and then multiply straight across.

Steps

1. Write the fractions in a single row separated by multiplication operators.
2. Multiply the numerators together.
3. Multiply the denominators together.

- $\frac{1}{2} \times \frac{3}{4} = \frac{3}{8}$

Tip: Fractions do not need a common denominator to be multiplied.

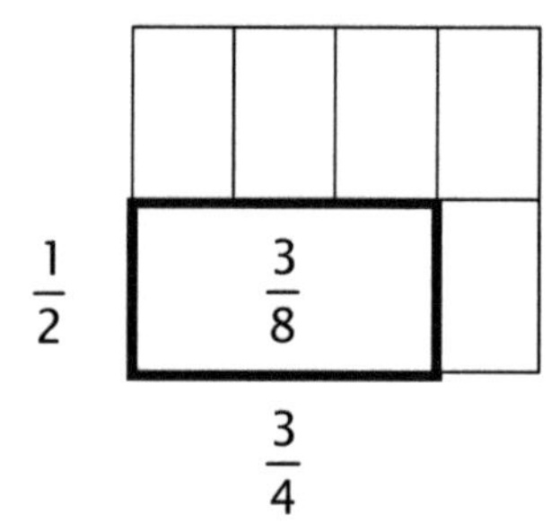

More than two fractions can be multiplied together at once.

- $\frac{1}{2} \times \frac{3}{10} \times \frac{1}{2} \times \frac{1}{4} = \frac{3}{160}$

Multiplying numbers that include fractions and whole numbers

When you multiply a fraction by a whole number, the whole number only multiplies the numerator of the fraction.

<u>Steps</u>

1. Write a 1 under each whole number to convert it to an improper fraction.

   Tip: This step is done just to make whole numbers look like fractions.

$$3 = 3 \div 1 = \frac{3}{1}$$

2. Multiply the numerators together.
3. Multiply the denominators together.

- $3 \times \frac{2}{7} = \frac{3}{1} \times \frac{2}{7} = \frac{6}{7}$

Multiplying numbers that include fractions and mixed numbers

When a multiplication problem includes mixed numbers, replace them with equivalent improper fractions and then multiply.

Steps

1. Replace each mixed number with an equivalent improper fraction.
2. Multiply the numerators together.
3. Multiply the denominators together.

- $\frac{2}{5} \times 1\frac{1}{3} = \frac{2}{5} \times \frac{4}{3} = \frac{8}{15}$

- $4\frac{1}{2} \times 2\frac{1}{3} = \frac{9}{2} \times \frac{7}{3} = \frac{63}{6} = \frac{21}{2} = 10\frac{1}{2}$

Tip: If you multiply the whole numbers together and multiply the fractions together, your answer will be wrong because it is too small.

$4\frac{1}{2} \times 2\frac{1}{3}$ equals $10\frac{1}{2}$ and not $8\frac{1}{6}$.

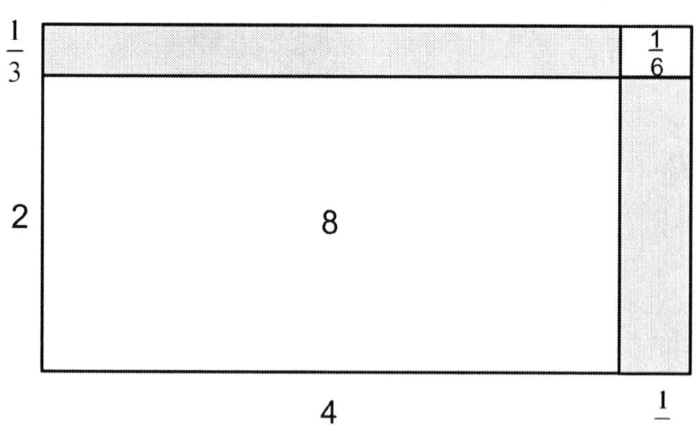

Writing the answer to a multiplication problem in simplest terms

Reduce a fraction that is not in lowest terms and replace an improper fraction with an equivalent whole number or mixed number.

- $4 \times \frac{2}{7} = \frac{4}{1} \times \frac{2}{7} = \frac{8}{7} = 1\frac{1}{7}$

- $2\frac{1}{4} \times 3 = \frac{9}{4} \times \frac{3}{1} = \frac{27}{4} = 6\frac{3}{4}$

- $1\frac{1}{2} \times 2\frac{1}{2} = \frac{3}{2} \times \frac{5}{2} = \frac{15}{4} = 3\frac{3}{4}$

- $\frac{3}{8} \times 1\frac{2}{3} \times 4 = \frac{3}{8} \times \frac{5}{3} \times \frac{4}{1} = \frac{60}{24} = 2\frac{12}{24} = 2\frac{1}{2}$

Canceling

Remove factors common to numerators and denominators before multiplying. This simplifies the problem and forces the answer to be in lowest terms.

- $\frac{1}{\cancel{2}_1} \times \frac{\cancel{6}^3}{5} = \frac{3}{5}$

  2 is divided into both a numerator and a denominator.

- $\frac{\cancel{3}^1}{\cancel{10}_2} \times \frac{\cancel{5}^1}{\cancel{6}_2} \times \frac{1}{3} = \frac{1}{12}$

  3 is canceled out of the 3 in the numerator and the 6 in the denominator. 5 is canceled out of the 5 and the 10.

NOTE: THE 1/3 REMAINS AS IT IS BECAUSE THERE IS NO OTHER MULTIPLE OF 3 IN A NUMERATOR TO CANCEL WITH IT.

Tip: If canceling weren't done before multiplying, reducing the answer might take some time.

$$\frac{3}{10} \times \frac{5}{6} \times \frac{1}{3} = \frac{15}{180} = ?$$

Reciprocal fractions

Two fractions, in which the numerator of each fraction equals the denominator of the other, are called RECIPROCALS of one another.

- $\frac{5}{2}$ and $\frac{2}{5}$ are reciprocals of one another.

- $\frac{7}{3}$ and $\frac{3}{7}$ are reciprocals of one another.

- $\frac{2}{3}$ and $1\frac{1}{2}$ are reciprocals, because $1\frac{1}{2} = \frac{3}{2}$.

- $\frac{1}{2}$ and 2 are reciprocals.

The product of two reciprocal fractions is 1.

- $\frac{5}{2} \times \frac{2}{5} = \frac{10}{10} = 1$

- $\frac{7}{3} \times \frac{3}{7} = \frac{21}{21} = 1$

- $\frac{2}{3} \times 1\frac{1}{2} = \frac{2}{3} \times \frac{3}{2} = \frac{6}{6} = 1$

- $\frac{1}{2} \times 2 = \frac{1}{2} \times \frac{2}{1} = \frac{2}{2} = 1$

Dividing Numbers That Include Fractions

Multiplying instead of dividing

When dividing numbers that include fractions, replace the divisor by its reciprocal and change the problem to one of multiplication. Then follow the procedure for multiplying fractions together.

Steps

1. Write the numbers in a single row as **(dividend) ÷ (divisor)**.

   "Divide 4 by $1\frac{1}{2}$." → $4 \div 1\frac{1}{2}$

2. Replace each whole and mixed number with an equivalent improper fraction.

   $$4 \div 1\frac{1}{2} = \frac{4}{1} \div \frac{3}{2}$$

3. Change the division operator (÷) to multiplication (×) and replace the divisor with its reciprocal.

   $$\frac{4}{1} \div \frac{3}{2} = \frac{4}{1} \times \frac{2}{3}$$

4. Multiply the fractions together.

   $$\frac{4}{1} \times \frac{2}{3} = \frac{8}{3}$$

5. Simplify the answer. Reduce a fraction that is not in lowest terms and replace an improper fraction with an equivalent whole number or mixed number.

   $$\frac{8}{3} = 2\frac{2}{3}$$

- $\frac{1}{5} \div 1\frac{1}{2} = \frac{1}{5} \div \frac{3}{2} = \frac{1}{5} \times \frac{2}{3} = \frac{2}{15}$
- $3 \div \frac{1}{2} = \frac{3}{1} \div \frac{1}{2} = \frac{3}{1} \times \frac{2}{1} = \frac{6}{1} = 6$

Tip: Do not cancel until the problem has been changed into a multiplication problem.

- In the following example, canceling a 2 out of 8 and 2 after the problem has been changed to one of multiplication will lead to the correct answer:

$$\frac{3}{8} \div \frac{1}{2} = \boxed{\frac{3}{8} \times \frac{2}{1}} = \boxed{\frac{3}{4} \times \frac{1}{1}} = \frac{3}{4} \; ☑$$

- In this example, canceling the 2s while it's still a division problem would give the wrong answer.

$$\boxed{\frac{2}{3} \div \frac{1}{2}} = \boxed{\frac{1}{3} \div \frac{1}{1}} = \frac{1}{3} \times \frac{1}{1} = \frac{1}{3} \; ☒$$

$$\frac{2}{3} \div \frac{1}{2} = \frac{2}{3} \times \frac{2}{1} = \frac{4}{3} = 1\frac{1}{3} \; ☑$$

- In this example, canceling 2 out of 4 and 6 while it's still a division problem would give the wrong answer.

$$1\frac{1}{3} \div \frac{1}{6} = \boxed{\frac{4}{3} \div \frac{1}{6}} = \boxed{\frac{2}{3} \div \frac{1}{3}} = \frac{2}{3} \times \frac{3}{1} = \frac{6}{3} = 2 \; ☒$$

A 3 can be canceled out of the 6 and 3 after it has been converted to a multiplication problem.

$$1\frac{1}{3} \div \frac{1}{6} = \frac{4}{3} \div \frac{1}{6} = \boxed{\frac{4}{3} \times \frac{6}{1}} = \boxed{\frac{4}{1} \times \frac{2}{1}} = \frac{8}{1} = 8 \; ☑$$

The procedure for dividing fractions also works with whole numbers. We don't use it to divide whole numbers because short and long division are much more efficient.

- We usually solve a division problem like this:

$$3\overline{)343}\quad 114\tfrac{1}{3}$$

But, we could solve the same problem using the procedure for dividing fractions.

$3\overline{)343}$ can be written as $343 \div 3$ which can be solved in a very round-about way:

$$343 \div 3 = \frac{343}{1} \div \frac{3}{1} = \frac{343}{1} \times \frac{1}{3} = \frac{343}{3} = 114\frac{1}{3}$$

Why multiply when dividing fractions

Instead of needing another procedure for dividing fractions, we reuse the one for multiplying fractions because it provides the correct answer.

- This example demonstrates why it works.
  1. A division problem is rewritten as a fraction.
  2. The fraction is multiplied by 1, whose numerator and denominator are written as the reciprocal of the divisor.
  3. The numerators and denominators of both fractions are then multiplied and simplified.

This demonstrates that dividing one fraction by another is the same as multiplying it by the reciprocal of the divisor.

$$\frac{2}{5} \div \frac{2}{3} = \frac{\frac{2}{5}}{\frac{2}{3}} = \frac{\frac{2}{5}}{\frac{2}{3}} \times 1 = \frac{\frac{2}{5}}{\frac{2}{3}} \times \frac{\frac{3}{2}}{\frac{3}{2}} = \frac{\frac{2}{5} \times \frac{3}{2}}{1} = \frac{2}{5} \times \frac{3}{2}$$

Complex fractions

Complex fractions are fractions that have a fraction in the numerator and/or the denominator.
Simplify a complex fraction by treating it as a division statement.

- $\dfrac{\frac{1}{4}}{2\frac{1}{2}} = \dfrac{1}{4} \div 2\dfrac{1}{2} = \dfrac{1}{4} \div \dfrac{5}{2} = \dfrac{1}{4} \times \dfrac{2}{5} = \dfrac{2}{20} = \dfrac{1}{10}$

Tip: When a fraction is divided by a fraction that has the same denominator, the complex fraction can be rapidly simplified by eliminating the denominators.

- $\dfrac{\frac{1}{4}}{\frac{3}{4}} = \dfrac{1}{3}$ because $\dfrac{\frac{1}{4}}{\frac{3}{4}} \times \dfrac{4}{4} = \dfrac{1}{3}$

Tip: A complex fraction that has a fraction in the numerator and a whole number in the denominator can be rapidly simplified by multiplying the denominator in the upper fraction by the whole number.

- $\dfrac{\frac{3}{2}}{7} = \dfrac{3}{14}$ because $\dfrac{\frac{3}{2}}{7} = \dfrac{3}{2} \div 7 = \dfrac{3}{2} \times \dfrac{1}{7} = \dfrac{3}{14}$

Comparing Operations on Fractions

<table>
<tr><th></th><th>Addition</th><th>Subtraction</th><th>Multiplication</th><th>Division</th></tr>
<tr><td></td><td>Two or more numbers</td><td>Only two numbers</td><td>Two or more numbers</td><td>Only two numbers</td></tr>
<tr><td rowspan="3">SETUP</td><td colspan="2">Write the numbers in columns.</td><td colspan="2">Write the numbers in a single row.</td></tr>
<tr><td colspan="2">DENOMINATORS MUST BE EQUAL.
If they are not, find the least common denominator. Then replace the fractions with equivalent fractions that have the common denominator.</td><td colspan="2">ALL TERMS MUST BE PROPER OR IMPROPER FRACTIONS.
If there are any mixed or whole numbers, write them as improper fractions.</td></tr>
<tr><td></td><td>THE BOTTOM FRACTION MUST BE LESS THAN THE TOP ONE.
If not, borrow 1 from the whole number on top.</td><td></td><td>Change "÷" to "×" and replace the divisor with its reciprocal (swap the numerator and denominator).</td></tr>
<tr><td rowspan="3">SOLVE</td><td></td><td></td><td colspan="2">Cancel common factors from the numerators and denominators.</td></tr>
<tr><td>Add numerators.
Copy the denominator.</td><td>Subtract numerators.
Copy the denominator.</td><td colspan="2">Multiply numerators.
Multiply denominators.</td></tr>
<tr><td>Add whole number terms.</td><td>Subtract whole number terms.</td><td colspan="2"></td></tr>
<tr><td>SIMPLIFY</td><td colspan="4">ANY FRACTION IN THE RESULT MUST BE IN LOWEST TERMS.
If needed, reduce the fraction to lowest terms.

ANY FRACTION IN THE RESULT MUST BE PROPER.
If the result has an improper fraction, change the improper fraction to a mixed or whole number.</td></tr>
</table>

Decimal Numbers

Decimal number. Rounding. Place values. Terminating decimal number. Non-terminating decimal number. Non-terminating, repeating decimal number. Non-terminating, non-repeating decimal number.

Representing Wholes and Parts as Decimal Numbers

A DECIMAL NUMBER is a number that has a decimal point separating a quantity of wholes from a quantity of parts.

The PLACE VALUES for decimal numbers increase by a multiple of 10. Each digit in a decimal number is a multiple of the place value it occupies.

Decimal place values are symmetric about the ones place and not symmetric about the decimal point. The decimal point merely separates the wholes from the parts. (There is no "oneths" place.)

Thousands	Hundreds	Tens	Ones	.	Tenths	Hundredths	Thousandths

- .4 Four tenths
- .03 Three hundredths
- .002 Two thousandths

Every whole number is a decimal number. Even though a whole number has no decimal point, the point can be added to the right of the ones column whenever whole numbers appear in a problem that includes numbers that have decimal points.

- 8 = 8.
- 604 = 604.

Decimal numbers are like mixed numbers because they can include both wholes and parts of a whole.

When reading a decimal numeral that includes a whole and parts of a whole, the decimal point is read as "and". This is the same as saying "and" when reading a mixed number.

- $5\frac{3}{10} = 5.3$ Five **and** three tenths
- $85\frac{2}{100} = 85.02$ Eighty-five **and** two hundredths

When reading the part in a decimal number, only the place value of the rightmost digit is mentioned.

- 92.648 is read as ninety-two and six hundred forty-eight **thousandths.** (The 8 is the rightmost digit and is in the thousandths place.)
- .67 = sixty-seven **hundredths**
  $.67 = \frac{6}{10} + \frac{7}{100} = \frac{60}{100} + \frac{7}{100} = \frac{67}{100}$
- .042 = forty-two **thousandths**
  $.042 = \frac{4}{100} + \frac{2}{1000} = \frac{40}{1000} + \frac{2}{1000} = \frac{42}{1000}$
- 8.301 = eight and three hundred one **thousandths**
  $8.301 = 8 + \frac{3}{10} + \frac{1}{1000} = 8 + \frac{300}{1000} + \frac{1}{1000}$
  $= 8\frac{301}{1000}$

Tip: Discard trailing zeros at the right of a decimal number.

- $.02\underline{0} = .02$ $(\frac{20}{1000} = \frac{2}{100})$
- $3.1\underline{000} = 3.1$ $(3\frac{1000}{10000} = 3\frac{1}{10})$

Rounding a Decimal Number to a Specific Place Value

The steps for rounding a decimal number are similar to those for rounding a whole number.

Steps

1. Begin by locating the digit in the place value to round to.
2. Focus on the digit to its right.
   - ONLY IF THE DIGIT TO ITS RIGHT IS 5 OR GREATER, add 1 to the digit in the place value you are rounding to.
3. Focus on whether the place value you are rounding to is to the left or right of the decimal point.
   - IF THE PLACE VALUE YOU ARE ROUNDING TO IS TO THE LEFT OF THE DECIMAL POINT, replace digits up to the decimal point with zeros and discard the decimal point and all digits to its right.
     - 192.8704
       - rounded to the nearest hundred is 200
       - rounded to the nearest ten is 190
       - rounded to the nearest whole number is 193
   - IF THE PLACE VALUE YOU ARE ROUNDING TO IS TO THE RIGHT OF THE DECIMAL POINT, discard all digits to its right.
     - 192.8704
       - rounded to the nearest tenth is 192.9
       - rounded to the nearest hundredth is 192.87
       - rounded to the nearest thousandth is 192.870

         (The 0 Is a SIGNIFICANT DIGIT and must be written, because the number is rounded to the thousandths place.)

Ordering Decimal Numbers

When ordering decimal numbers it helps to give them each the same number of digits to the right of the decimal point by writing additional zeros if necessary.

- To order 7.099, 7.5, and 7.20:

  Add zeros to 7.5 and 7.20. (This makes every number have three digits to the right of the decimal point.)

  7.099, 7.500, 7.200

  Write these numbers in ascending order:

  7.099, 7.200, 7.500

  Then write the original numbers in order:

  7.099, 7.20, 7.5

Types of Decimal Numbers

When a decimal number has a final digit it is called a TERMINATING decimal number.

- .2
- .375
- 6.5

When the digits in a decimal number go on forever, it is called a NON-TERMINATING decimal number.

Three dots (...) are written after it to indicate that it goes on forever.

- .333...
- 1.486...

When the digits in a non-terminating decimal number repeat, it is called a NON-TERMINATING, REPEATING decimal number.

A line is written over the repeating digits to indicate that they repeat.

- $6.333... = 6.\overline{3}$
- .28571428571428571428571428571428...

  $= .2857\overline{142857142857}$

When the digits in a non-terminating decimal number do not repeat, it is called a NON-TERMINATING, NON-REPEATING decimal number.

- .4142135623730950488016887242097...
- 1.7320508075688772935274463415059...

Adding Decimal Numbers

The most important thing to do when adding decimal numbers is to line up the decimal points.

Steps

1. Write a decimal point to the right of any whole number.
2. Write the numbers beneath one another with their decimal points lined up. (This forces the place values to line up.)
3. Fill-in zeros to make each number have the same number of places to the right of the decimal point.
4. Add the numbers as if there were no decimal points.
5. Place a decimal point in the answer under the other decimal points.

- $38.5 + 6 + 2.73 = 47.23$

$$\begin{array}{r} 38.50 \\ 6.00 \\ \underline{+\ 2.73} \\ 47.23 \end{array}$$

- $.24 + 2.4 + 24 = 26.64$

$$\begin{array}{r} .24 \\ 2.40 \\ \underline{+\ 24.00} \\ 26.64 \end{array}$$

Subtracting Decimal Numbers

The most important thing to do when subtracting decimal numbers is to line up the decimal points.

Steps

1. Write a decimal point to the right of any whole number.

2. Write the smaller number beneath the larger number with their decimal points lined up. (This forces the place values to line up.)

3. Fill-in zeros to make each number have the same number of places to the right of the decimal point.

4. Subtract the numbers as if there were no decimal points.

5. Place a decimal point in the answer under the other decimal points.

- $8 - .37 = 7.63$

$$\begin{array}{r} 8.00 \\ \underline{-\ .37} \\ 7.63 \end{array}$$

- $94.2 - 6.34 = 87.86$

$$\begin{array}{r} 94.20 \\ \underline{-\ \ 6.34} \\ 87.86 \end{array}$$

Multiplying Decimal Numbers

The most important thing to do when multiplying decimal numbers is to place the decimal point correctly in the answer.

Steps

1. Write the numbers beneath one another with their rightmost digits beneath one another.

$$\begin{array}{r} .222 \\ \underline{\times\ 1.3} \end{array}$$

2. Multiply as if there were no decimal point.

$$\begin{array}{r} .222 \\ \underline{\times\ 1.3} \\ 666 \\ \underline{222} \\ 2886 \end{array}$$

3. When you get the result, count the number of digits to the right of the decimal points in the numbers that were multiplied.

   There are four digits to the right of the decimal points: .222 and .3

4. Place the decimal point in the answer so that the answer has the same number of digits to the right of the point as the count you made in Step 3.

$$\begin{array}{r} .222 \\ \underline{\times\ 1.3} \\ 666 \\ \underline{222} \\ .2886 \end{array}$$

- $$\begin{array}{r} .06 \\ \times\ .2 \\ \hline .012 \end{array}$$

  There are two digits to the right of the point in .06 and one digit to the right of the point in .2 (Since 2 + 1 = 3, the decimal point is placed in the answer so that three digits are to its right.)

  NOTE: REPLACING THE DECIMAL NUMBERS WITH FRACTIONS ILLUSTRATES WHY THIS WORKS.

  $$.2 \times .06 = \frac{2}{10} \times \frac{6}{100} = \frac{12}{1000} = .012$$

- $$\begin{array}{r} 1234.2 \\ \times\ .02 \\ \hline 24.684 \end{array}$$

- $12.3 \times .1 = 1.23$
- $12.3 \times .01 = .123$
- $12.3 \times .001 = .0123$

If one of the numbers being multiplied is a whole number, it will not have a decimal point. The implied decimal point is to its right and there are no digits following it.

- $$\begin{array}{r} .12345 \\ \times\ 22 \\ \hline 24690 \\ 24690 \\ \hline 2.71590 \end{array}$$

  22 has no decimal point. The point is placed in the answer so that there are five digits to its right.

Dividing Decimal Numbers

To divide decimal numbers, focus on the decimal points in the divisor, the dividend, and the quotient.

- $.11\overline{)2.222}$ — Write the division problem using the box-like division symbol.

  $11.\overline{)2.222}$ — Make the divisor a whole number by moving its decimal point to the extreme right.

  $11.\overline{)222.2}$ — Move the decimal point of the dividend the same number of digits to the right.

  $\begin{array}{r} . \\ 11.\overline{)222.2} \end{array}$ — Put a decimal point in the answer directly above the decimal point in the dividend.

  $\begin{array}{r} 20.2 \\ 11.\overline{)222.2} \\ -\underline{22} \\ 22 \\ \underline{-22} \end{array}$ — Divide 11 into 222.2, writing digits in correct columns.

NOTE: THE ZERO WAS WRITTEN IN THIS ANSWER BECAUSE 11 COULD NOT DIVIDE THE FIRST 2 THAT WAS BROUGHT DOWN.

$$\begin{array}{r} 20. \\ 11.\overline{)222.2} \\ -\underline{22}\downarrow \\ 2 \end{array}$$

NOTE: WE MOVE THE DECIMAL POINT IN THE DIVISOR AND THE DIVIDEND BEFORE WE START TO DIVIDE IN ORDER TO CORRECTLY PLACE THE DECIMAL POINT IN THE QUOTIENT.

$$\begin{array}{r} 2 \\ .03\overline{)6} \end{array} \boxtimes \qquad .03\overline{)6} = \begin{array}{r} 200. \\ 3\overline{)600.} \end{array} \ \boxed{\checkmark}$$

Verifying with fractions: $6 \div .03 = 6 \div \frac{3}{100} = \frac{6}{1} \times \frac{100}{3} = 200$

Don't stop if the answer has a remainder. Add more zeros to the dividend until your answer has one digit more than the place value you are rounding to.

- Dividing 2.1 by 5 to the nearest tenth:

$$5\overline{)2.1}\quad \text{quotient } .4$$

20
1 — There is a remainder.

$$5\overline{)2.10}\quad \text{quotient } .42$$

Add a 0 and continue dividing.

20
10
10

To the nearest tenth, 2.1 ÷ 5 = .4

- Dividing 30 by 7 to the nearest hundredth:

$$7\overline{)30.000}\quad \text{quotient } 4.285$$

28
20
14
60
56
40
35
5

To the nearest hundredth, 30 ÷ 7 = 4.29

Decimal Numbers and Equivalent Fractions

Converting a fraction to a decimal number

To convert a proper fraction to a decimal number, divide the denominator into the numerator.

- To write $\frac{3}{8}$ as a decimal number, divide 8 into 3. $8\overline{)3.000}$ = .375 $\frac{3}{8} = .375$

- To write $\frac{1}{2}$ as a decimal number, divide 2 into 1. $2\overline{)1.0}$ = .5 $\frac{1}{2} = .5$

- To write $\frac{1}{12}$ as a decimal number, divide 1 by 12. $1 \div 12 = .08333...$ $\frac{1}{12} = .08\overline{3}$

Converting a mixed number to a decimal number

To convert a mixed number to a decimal number, copy the whole number that is in the mixed number, and then convert the fraction to a decimal number.

- $1\frac{3}{8} = 1.375$
- $4\frac{1}{2} = 4.5$
- $8\frac{3}{4} = 8.75$
- $7\frac{1}{12} = 7.08\overline{3}$

Converting a terminating decimal number to a fraction

The words we say when properly reading a decimal number can simply be written as a fraction or a mixed number.

- $.3 =$ "Three tenths" $= \frac{3}{10}$
- $.014 =$ "Fourteen thousandths" $= \frac{14}{1000} = \frac{7}{500}$
- $5.36 =$ "Five and thirty-six hundredths"

  $= 5\frac{36}{100} = 5\frac{9}{25}$

Converting a repeating decimal number to a fraction

To convert a non-terminating, repeating decimal number to a fraction, place the repeating digits over just as many 9s.

- $.\overline{48} = \frac{48}{99} = \frac{16}{33}$
- $.\bar{3} = \frac{3}{9} = \frac{1}{3}$
- $8.\overline{001} = 8\frac{1}{999}$

To convert a repeating decimal that starts with non-repeating digits, first isolate the repeating digits.

- $$\begin{aligned} .7\overline{14} &= .7 + .01414 \\ &= .7 + (.1)(.1414...) \\ &= \frac{7}{10} + \frac{1}{10}(\frac{14}{99}) = \frac{7}{10} + \frac{14}{990} \\ &= \frac{7(99)}{990} + \frac{14}{990} = \frac{693}{990} + \frac{14}{990} = \frac{707}{990} \end{aligned}$$

NOTE: NON-TERMINATING, NON-REPEATING DECIMAL NUMBERS CANNOT BE WRITTEN AS FRACTIONS.

Ordering fractions and decimal numbers

To order numbers that include both fractions and decimal numbers, first convert the fractions to decimal numbers, and then order the decimal numbers.

- To determine the greater of $\frac{1}{8}$ and .2, first convert $\frac{1}{8}$ to a decimal number. $\frac{1}{8} = .125$

  .2 is greater than $\frac{1}{8}$ because .200 is greater than .125

Universal Number Concepts

Set notation { }. Power. Square. Square root. Radical symbol $\sqrt{\ }$. Cube. Cube root. Exponential notation. Exponential. Base. Exponent. Value of an exponential. Power of a power. Prime factor. Exponent of 0. Fractional exponent. Scientific notation. Order of operations. PEMDAS. Commutative, associative, and distributive properties. Inequality symbols. Not equal to symbol ≠. Less than symbol <. Greater than symbol >. Less than or equal to symbol ≤. Greater than or equal to symbol ≥. Number line. Graphing numbers.

NOTE: THE TOPICS IN THIS SECTION DESCRIBE CONCEPTS THAT APPLY TO ALL NUMBERS, INCLUDING WHOLE NUMBERS, FRACTIONS, DECIMAL NUMBERS, AND POSITIVE AND NEGATIVE NUMBERS.

Set Notation

A notation is a way of writing something in a standard way. Set notation uses curly braces { } to surround the group of items that make up a set.

- {1, 2, 3, ...} is a way of describing all of the natural numbers.
- {0, 1, 2, 3, ...} is a way of describing all of the whole numbers.
- {2, 3} is a way of describing the whole numbers that are between 1 and 4.
- $\{\frac{1}{4}, \frac{1}{2}, \frac{3}{4}, 1, 1\frac{1}{4}, 1\frac{1}{2}, 1\frac{3}{4}\}$ is a way of describing the numbers between 0 and 2 that are multiples of $\frac{1}{4}$.
- {0, .2, .4, .6, ...} is a way of describing the decimal numbers that increase by .2 beginning with zero.

Powers

Powers of a number

When several identical numbers are multiplied together the result is called a POWER of that number.

- $10 = 10$

  $10 \times 10 = 100$

  $10 \times 10 \times 10 = 1000$

  10, 100, and 1000 are powers of 10.

- $2 = 2$

  $2 \times 2 = 4$

  $2 \times 2 \times 2 = 4 \times 2 = 8$

  $2 \times 2 \times 2 \times 2 = 8 \times 2 = 16$

  $2 \times 2 \times 2 \times 2 \times 2 = 16 \times 2 = 32$

  $2 \times 2 \times 2 \times 2 \times 2 \times 2 = 32 \times 2 = 64$

  2, 4, 8, 16, 32, and 64 are powers of 2.

The same number could be a power of more than one number.

- $64 = 64$
- $64 = 8 \times 8$
- $64 = 4 \times 4 \times 4$
- $64 = 2 \times 2 \times 2 \times 2 \times 2 \times 2$

64 is a power of 64, 8, 4 and 2.

There are several ways to verbally express the power of a number.

- The following phrases are commonly used to express $2 \times 2 \times 2 \times 2$.

  "Two raised to the fourth power"
  "Two to the fourth power"
  "Two to the fourth"

The NUMBER OF A POWER identifies how many base numbers are multiplied together.

- $4 = 2 \times 2$ — 4 is the **2nd power** of 2.
- $8 = 2 \times 2 \times 2$ — 8 is the **3rd power** of 2.
- $6 = 6$ — 6 is the **1st power** of 6.
- $64 = 8 \times 8$ — 64 is the **2nd power** of 8
- $64 = 4 \times 4 \times 4$ — 64 is the **3rd power** of 4
- $64 = 2 \times 2 \times 2 \times 2 \times 2 \times 2$ — 64 is the **6th power** of 2.

The number of a power of 10 is equal to the number of zeros in the power of 10.

- 10 is the first power of ten.
- 100 is the second power of ten.
- 1000 is the third power of ten.

Squares, square roots, cubes, and cube roots

The 2nd power of a number is called its SQUARE.

- $3 \times 3 = 9$ — 3 squared is 9; 9 is the square of 3
- $6 \times 6 = 36$ — 6 squared is 36; 36 is the square of 6
- $8 \times 8 = 64$ — 8 squared is 64; 64 is the square of 8

The number that is multiplied by itself to result in a square is called the SQUARE ROOT of the result. The notation for square root is a RADICAL SYMBOL $\sqrt{\ }$.

- $3 \times 3 = 9$ — The square root of 9 is 3. — $\sqrt{9} = 3$
- $6 \times 6 = 36$ — The square root of 36 is 6. — $\sqrt{36} = 6$
- $8 \times 8 = 64$ — 8 is the square root of 64. — $\sqrt{64} = 8$

Most square roots are non-repeating, non-terminating decimal numbers.

- $\sqrt{2} = 1.4142135623730950488016887242097...$
- $\sqrt{3} = 1.7320508075688772935274463415059...$
- $\sqrt{5} = 2.2360679774997896964091736687313...$
- $\sqrt{6} = 2.4494897427831780981972840747059$
- $\sqrt{7} = 2.6457513110645905905016157536393...$

The square root of a decimal number can be calculated with pencil and paper.

- To find $\sqrt{23.2}$ to the nearest tenth:

```
    4 .
  √23 . 20 00
   -16
     7
```

1. Pair off the digits on both sides of the decimal point. (Add enough zeros to so that there are four digits to the right of the decimal point – two for the tenths place and two for the hundredths place.)
2. Beneath 23, write the largest whole number square that is less than 23, which is 16.
3. Write the square root of 16 above 23.
4. Subtract 16 from 23.

```
        4 .
      √23 . 20 00
       -16
 8__ |   7  20
```

5. Bring down the next pair of digits.
6. Double the 4 and write 8 to the left of the bottom row. Leave a space for another digit.

```
        4 . 8
      √23 . 20 00
       -16
 88  |   7  20
        -7  04
            16
```

7. Determine the largest digit such that ? x 8<u>?</u> is less than 720. (7 x 87 = 609 is too small and 9 x 89 = 801 is too large.) Write an 8 in both places.
8. Multiply **8** x 8**8**, write 704 beneath 720, and subtract.

```
        4 . 8
      √23 . 20 00
       -16
 88  |   7  20
        -7  04
 96_ |      16 00
```

9. Bring down the next pair of digits.
10. Double the 48 and write 96 to the left of the bottom row. Leave a space for another digit.

```
        4 . 8  1
      √23 . 20 00
       -16
 88  |   7  20
        -7  04
 961 |      16 00
            -9 61
             6 39
```

11. Determine the largest digit such that ? x 96<u>?</u> is less than 1600. (2 x 962 = 1924 is too large.) Write a 1 in both places.
12. Multiply **1** x 96**1**, write 961 beneath 1600, and subtract.

To the nearest tenth, $\sqrt{23.2}$ = 4.8

CHECK: 4.8 X 4.8 = 23.04

Tip: If there is only one digit in the left-most pair, begin the same way.

```
       2.
     √ 5.0000
      -4
 4_ |  100
```

The 3rd power of a number is called its CUBE.

- $1 \times 1 \times 1 = 1$ 1 cubed is 1.
  1 is the cube of 1.
- $2 \times 2 \times 2 = 8$ 2 cubed is 8.
  8 is the cube of 2.
- $3 \times 3 \times 3 = 27$ 3 cubed is 27.
  27 is the cube of 3.

When three identical numbers are multiplied to result in a cube, the number is called the CUBE ROOT of the result. The notation for cube root is a radical symbol preceded by a small 3.

- $1 \times 1 \times 1 = 1$ The cube root of 1 is 1. $\sqrt[3]{1} = 1$
- $2 \times 2 \times 2 = 8$ The cube root of 8 is 2. $\sqrt[3]{8} = 2$
- $3 \times 3 \times 3 = 27$ 3 is the cube root of 27. $\sqrt[3]{27} = 3$

Multiplying by powers of 10

To multiply a decimal number by a power of 10, move the decimal point to the right as many places as there are zeros in the power of 10.

When the point must move past a place that doesn't have a digit, write a zero in that place.

NOTE: MOVING THE POINT TO THE RIGHT MAKES THE DECIMAL NUMBER LARGER.

- $12 \times 10 = 120$
- $2.5 \times 10 = 25$
- $462.04 \times 10 = 4{,}620.4$

10 has **one** zero. To multiply any decimal number by 10, move the decimal point **one** place to the right.

- $5 \times 100 = 500$

100 has **two** zeros. To multiply any decimal number by 100, move the decimal point **two** places to the right.

- $.7 \times 1000 = 700$

1000 has **three** zeros. To multiply any decimal number by 1000, move the decimal point **three** places to the right.

Dividing by powers of 10

To divide a decimal number by a power of 10, move the decimal point to the left as many places as there are zeros in the power of 10.

When the point must move past a place that doesn't have a digit, write a zero in that place.

NOTE: MOVING THE POINT TO THE LEFT MAKES THE DECIMAL NUMBER SMALLER.

- $2 \div 10 = .2$
- $10 \div 10 = 1$
- $2.5 \div 10 = .25$
- $462.04 \div 10 = 46.204$

10 has **one** zero in it. To divide any decimal number by 10, move the decimal point **one** place to the left.

- $5 \div 100 = .05$

100 has **two** zeros in it. To divide any decimal number by 100, move the decimal point **two** places to the left.

- $.7 \div 1000 = .0007$

1000 has **three** zeros in it. To divide any decimal number by 1000, move the decimal point **three** places to the left.

Exponential Notation

Exponentials

Exponential notation is a simplified way to write a power of a number.

- 2^7 is an example of a power written in exponential notation.

  $2^7 = 2 \times 2 \times 2 \times 2 \times 2 \times 2 \times 2$

Each symbol used in exponential notation has a name.

- 2^7 is called an EXPONENTIAL.

  2 is the BASE of the exponential.

  7 is the EXPONENT.

To find the VALUE OF AN EXPONENTIAL, multiply as many base numbers together as indicated by the exponent.

- The value of $2^4 = 2 \times 2 \times 2 \times 2 = 16$
- The value of $30^2 = 30 \times 30 = 900$
- The value of $5^3 = 5 \times 5 \times 5 = 125$

An exponent of 1 indicates that the value of the exponential equals its base.

- $5^1 = 5$
- $23^1 = 23$
- $2^1 = 2$

When no exponent is written, the implied exponent is 1.

- $5 = 5^1$
- $23 = 23^1$
- $2 = 2^1$

Simplifying exponential products and quotients

Simplify the product or quotient of exponentials whenever the numeric value of the product does not have to be precisely known, or when finding the value would require excessive calculation.

Add exponents to simplify the product of exponentials that have the same base.

- $5^3 \times 5^5 = 5^8$

$(5 \cdot 5 \cdot 5) \times (5 \cdot 5 \cdot 5 \cdot 5 \cdot 5)$

$= 5 \cdot 5 \cdot 5 \cdot 5 \cdot 5 \cdot 5 \cdot 5 \cdot 5$

Multiplying three 5s by five 5s is the same as multiplying 8 fives together, so $5^3 \cdot 5^5 = 5^8$.

- $10^4 \times 10^3 = 10^7$
- $3^6 \times 3^2 = 3^8$
- $2^5 \times 2^3 = 2^8$
- $1^{10} \times 1^6 = 1^{16}$
- $8^3 \times 8^3 \times 8^3 = 8^9$
- $(5^7)(5^3)(5^6) = 5^{16}$
- $4^2 \times 4^2 \times 4^2 \times 4^2 = 4^8$
- $2 \times 2^3 = 2^1 \times 2^3 = 2^4$

NOTE: THE RESULT HAS THE SAME BASE AS THE EXPONENTIALS BEING MULTIPLIED.

Tip: Do not add or multiply the bases.

To simplify a POWER OF A POWER, multiply exponents.

- $(10^2)^3 = (10^2)\,(10^2)\,(10^2) = 10^6$ You can either add 2+2+2 or multiply 2×3 to get 6.

Subtract exponents to simplify the quotient of exponentials that have the same base.

- $\dfrac{10^5}{10^3} = 10^2$ Three 10s can be canceled from the top and bottom:

  $$\frac{\cancel{10}\times\cancel{10}\times\cancel{10}\times 10\times 10}{\cancel{10}\times\cancel{10}\times\cancel{10}} = 10 \times 10 = 10^2$$

- $\dfrac{8^7}{8^4} = 8^3$ Four 8s can be canceled from the top and bottom:

  $$\frac{\cancel{8}\times\cancel{8}\times\cancel{8}\times\cancel{8}\times 8\times 8\times 8}{\cancel{8}\times\cancel{8}\times\cancel{8}\times\cancel{8}} = 8 \times 8 \times 8 = 8^3$$

- $\dfrac{4^3}{4^1} = 4^2$
- $\dfrac{5^5}{5} = \dfrac{5^5}{5^1} = 5^4$
- $10^6 \div 10^2 = 10^4$

NOTE: THE RESULT HAS THE SAME BASE AS THE EXPONENTIALS BEING DIVIDED.

Tip: Do not subtract or divide the bases.

Writing prime factors in exponential notation

Because a composite number often contains several identical prime factors, exponential notation is commonly used to describe the prime factors of a whole number.

- $12 = 2 \times 2 \times 3 = 2^2 \cdot 3$
- $8 = 2 \times 2 \times 2 = 2^3$
- $100 = 4 \cdot 25 = 2 \cdot 2 \cdot 5 \cdot 5 = 2^2 \cdot 5^2$

Multiplying and dividing by powers of 10 written in exponential notation

To multiply a decimal number by a power of 10 written in exponential notation, move the decimal point to the right as many places as the exponent of the power of 10.

- $83 \times 10^1 = 830$ $(10^1 = 10)$
- $2.5 \times 10^2 = 250$ $(10^2 = 100)$
- $.0012 \times 10^3 = 1.2$ $(10^3 = 1000)$

To divide a decimal number by a power of 10 written in exponential notation, move the decimal point to the left as many places as the exponent of the power of 10.

- $720 \div 10^2 = 7.2$
- $.31 \div 10^4 = .000031$
- $\dfrac{2.5}{10^3} = .0025$

Scientific notation (for large numbers)

Scientific notation is the way scientists write approximations for very large numbers.

Scientific notation is the product of two terms.
The first term is a number between 1 and 10.
The second term is a power of 10.

- 3.475×10^7 To find the value of this number, move the point 7 places to the right.

  $3.475 \times 10^7 = 34{,}750{,}000.$

- 8.0×10^{15} To find the value of this number, move the point 15 places to the right.

  $8.0 \times 10^{15} = 8{,}000{,}000{,}000{,}000{,}000.$

Any large number can be written in scientific notation.

Steps

1. Copy the digits, ignoring the zeros at the end of the number.

   497,000,000,000 = **497**

2. Place a decimal point after the first digit in the number that you wrote.

   497,000,000,000 = **4.97**

   (4.97 is between 1 and 10.)

3. Skip the first digit in the original number and count the remaining digits.

   There are 11 digits in: 97,000,000,000

4. Use this number as the exponent in the power of 10.

   497,000,000,000 = $\mathbf{4.97 \times 10^{11}}$

NOTE: MULTIPLYING LARGE NUMBERS WITH A CALCULATOR CAN DISPLAY AN ANSWER SIMILAR TO: `2.5 E15.`

THE **E** REPRESENTS "EXPONENT". `2.5 E15` $= 2.5 \times 10^{15}$

An exponent of 0

Zero can be an exponent.

When simplifying a quotient of exponentials that have equal bases, exponents are subtracted.

When the exponents are equal the result will have an exponent of zero.

$$\frac{2^3}{2^3} = 2^0$$

An exponent of 0 indicates that the *value* of the exponential is 1.

- Since $\frac{2^3}{2^3} = 2^0$ and $\frac{2^3}{2^3} = \frac{8}{8} = 1$, then $2^0 = 1$,

The same is true for any base:

- $5^0 = 1$
- $8^0 = 1$

Simplify as usual:

- $5^3 \times 5^0 = 5^3$ To verify this result, replace 5^0 by 1:

  $5^3 \times 1 = 5^3$

- $\frac{2^5}{2^0} = 2^5$ To verify this result, replace 2^0 by 1:

  $2^5 \div 1 = 2^5$

Tip: Having an exponent of 0 is different from having no exponent. When there is no exponent, the exponent is really 1, and the value of the exponential is its base.

Fractional exponents

An exponent of $\frac{1}{2}$ indicates that the exponential represents a square root.

Because we add exponents to simplify the product of exponentials that have the same base:

$$5^{\frac{1}{2}} \times 5^{\frac{1}{2}} = 5^1$$

But, any number that is multiplied by itself is the square root of the result.

$$\sqrt{5} \times \sqrt{5} = 5$$

So, $5^{\frac{1}{2}}$ must equal $\sqrt{5}$.

The same is true for any base.

- $36^{\frac{1}{2}} = \sqrt{36} = 6$
- $9^{\frac{1}{2}} = 3$
- $25^{\frac{1}{2}} = 5$

Powers of fractions and decimal numbers

Powers of fractions can be written in exponential notation.

- $\left(\frac{1}{2}\right)^2 = \frac{1}{2} \times \frac{1}{2} = \frac{1}{4}$

  $\frac{1}{2}$ squared equals $\frac{1}{4}$

  $\frac{1}{4}$ is the square of $\frac{1}{2}$

- $\left(\frac{2}{3}\right)^3 = \frac{2}{3} \times \frac{2}{3} \times \frac{2}{3} = \frac{8}{27}$

  $\frac{2}{3}$ cubed equals $\frac{8}{27}$

  $\frac{8}{27}$ is the cube of $\frac{2}{3}$

- $(\frac{1}{2})^4 = \frac{1}{2} \times \frac{1}{2} \times \frac{1}{2} \times \frac{1}{2} = \frac{1}{16}$

- $\left(\frac{1}{2}\right)^3 \times \left(\frac{1}{2}\right)^2 = \left(\frac{1}{2}\right)^5$

- $\dfrac{\left(\frac{1}{2}\right)^5}{\left(\frac{1}{2}\right)^2} = \left(\frac{1}{2}\right)^3$

- $(\frac{1}{2})^1 = \frac{1}{2}$

- $(\frac{1}{2})^0 = 1$

Powers of decimal numbers can be written in exponential notation.

- $(.5)^2 = .5 \times .5 = .25$ — .5 squared equals .25
  .25 is the square of .5
- $(.5)^3 = .5 \times .5 \times .5 = .125$ — .5 cubed equals .125
  .125 is the cube of .5
- $(.5)^4 = .5 \times .5 \times .5 \times .5 = .0625$
- $(.5)^2 \times (.5)^3 = (.5)^5$
- $(.5)^5 \div (.5)^3 = (.5)^2$
- $(.5)^1 = .5$
- $(.5)^0 = 1$

NOTE: A POWER OF A NUMBER CAN BE LESS THAN ITS BASE.

- $8^0 = 1$ — 1 is less than 8
- $(.5)^2 = .5 \times .5 = .25$ — .25 is less than .5
- $\left(\frac{2}{3}\right)^3 = \frac{2}{3} \times \frac{2}{3} \times \frac{2}{3} = \frac{8}{27}$ — $\frac{8}{27}$ is less than $\frac{2}{3}$

Order of Operations

Operations that should be performed together are grouped by parentheses. Operations within parentheses should be performed before other operations.

The following operations on identical numbers are grouped differently and yield different results.

- $(8 + 4) \times 2 \quad = (12) \times 2 = 24$

  $8 + (4 \times 2) \quad = 8 + (8) = 16$

- $\frac{1}{4} \times (2 + \frac{1}{4}) \quad = \frac{1}{4} \times (2\frac{1}{4}) = \frac{1}{4} \times \frac{9}{4} = \frac{9}{16}$

  $(\frac{1}{4} \times 2) + \frac{1}{4} \quad = (\frac{2}{4}) + \frac{1}{4} = \frac{3}{4}$

- $(60 - .6) \div 2 \quad = (59.4) \div 2 = 29.7$

  $60 - (.6 \div 2) \quad = 60 - (.3) = 59.7$

To simplify expressions consisting of several numbers interspersed with operators and parentheses, a specific order of operations must be followed.

Steps

1. **Simplify expressions within PARENTHESES**

   Parentheses are usually written as: ( ).
   IF A PARENTHESIS IS WRITTEN INSIDE ANOTHER PARENTHESIS, the outer one is written as [ ] or { }.

   First simplify the numbers in the inside parenthesis, then simplify the numbers in the outer parenthesis.

When all expressions within parentheses have been simplified:

2. **Replace EXPONENTIALS with their values.**

When all exponential numbers have been replaced by their values:

3. **MULTIPLY and DIVIDE**

   Move left to right, performing multiplications and divisions as they are encountered.

When the only remaining operations are addition and subtraction:

4. **ADD and SUBTRACT**

   Move left to right, performing additions and subtractions as they are encountered.

Tip: A common acronym used to remember the order of operations is PEMDAS.
("Please Excuse My Dear Aunt Sally.")

- $2\{ 50 - [2(2 + 3)] \}$ First simplify $(2 + 3)$.
  $= 2\{ 50 - [2(5)] \}$ Then $[2(5)]$.
  $= 2\{ 50 - 10 \}$ Then $\{50 - 10\}$.
  $= 2\{40\}$ Then $2\{40\}$.
  $= 80$

- $20 - 2(1 + 1)^3$ First simplify $(1 + 1)$.
  $= 20 - 2(2)^3$ Then 2^3.
  $= 20 - 2(8)$ Then $2(8)$.
  $= 20 - 16 = 4$ Then $20 - 16$.

- $8 + 4 \times 21 - 12 \div 6 - 4$ First simplify 4×21.
  $= 8 + 84 - 12 \div 6 - 4$ Then $12 \div 6$.
  $= 8 + 84 - 2 - 4$ Then $8 + 84$.
  $= 92 - 2 - 4$ Then $92 - 2$.
  $= 90 - 4 = 86$ Then $90 - 4$.

- $8 + 4 (21 - 12) \div 6 - 4$ First simplify $(21 - 12)$.
  $= 8 + 4(9) \div 6 - 4$ Then $4(9)$.
  $= 8 + 36 \div 6 - 4$ Then $36 \div 6$.
  $= 8 + 6 - 4$ Then $8 + 6$.
  $= 14 - 4 = 10$ Then $14 - 4$.

- $8 + 4 (21 - 12) \div (6 - 4)$ First simplify $(21 - 12)$ and $(6 - 4)$.
  $= 8 + 4(9) \div 2$ Then $4(9)$.
  $= 8 + 36 \div 2$ Then $36 \div 2$.
  $= 8 + 18 = 26$ Then $8 + 18$.

Properties of Numeric Operations

Commutative property

When the position of numbers can be changed in an expression without changing the result of an operation, the operation is said to have a commutative property.

Addition and multiplication both have commutative properties.

The commutative property of addition:

- $5 + 2 = 2 + 5$
- $\frac{1}{2} + \frac{2}{3} = \frac{2}{3} + \frac{1}{2}$
- $1.5 + .5 = .5 + 1.5$

The commutative property of multiplication:

- $2 \times 3 = 3 \times 2$
- $\frac{3}{4} \times \frac{1}{4} = \frac{1}{4} \times \frac{3}{4}$
- $6.1 \times 2 = 2 \times 6.1$

Associative property

When the order in which operations are performed does not affect the final result, the operation is said to have an associative property.

Addition and multiplication both have associative properties.

The associative property of addition:

- $2 + (3 + 4) = (2 + 3) + 4$
  $2 + 7 = 5 + 4$

- $\frac{1}{2} + (1 + \frac{1}{4}) = (\frac{1}{2} + 1) + \frac{1}{4}$
  $\frac{1}{2} + 1\frac{1}{4} = 1\frac{1}{2} + \frac{1}{4}$

- $.5 + (.1 + .2) = (.5 + .1) + .2$
  $.5 + .3 = .6 + .2$

The associative property of multiplication:

- $2 \times (3 \times 4) = (2 \times 3) \times 4$
  $2 \times 12 = 6 \times 4$

- $\frac{1}{2} \times (\frac{1}{4} \times \frac{1}{3}) = (\frac{1}{2} \times \frac{1}{4}) \times \frac{1}{3}$
  $\frac{1}{2} \times \frac{1}{12} = \frac{1}{8} \times \frac{1}{3}$

- $2.1 \times (5 \times .6) = (2.1 \times 5) \times .6$
  $2.1 \times 3 = 10.5 \times .6$

Distributive property

When an operation can be distributed over all the terms within a parenthesis, it is said to have a distributive property.

Multiplication is distributive over addition and subtraction.

The distributive property of multiplication over addition:

- $2(3 + 4) = (2 \times 3) + (2 \times 4)$

  Simplifying the terms on each side of the equals sign shows that both sides are equal.

  $2(3 + 4) = 2(7) = 14$

  $(2 \times 3) + (2 \times 4) = 6 + 8 = 14$

- $\frac{2}{3}(3 + \frac{1}{2}) = (\frac{2}{3} \times 3) + (\frac{2}{3} \times \frac{1}{2})$

  Simplifying the terms on each side of the equals sign shows that both sides are equal.

  $\frac{2}{3}(3 + \frac{1}{2}) = \frac{2}{3} \times (3\frac{1}{2}) = \frac{2}{3} \times \frac{7}{2} = \frac{7}{3} = 2\frac{1}{3}$

  $(\frac{2}{3} \times 3) + (\frac{2}{3} \times \frac{1}{2}) = 2 + \frac{1}{3} = 2\frac{1}{3}$

- $10(.3 + .4) = (10 \times .3) + (10 \times .4)$

  Simplifying the terms on each side of the equals sign shows that both sides are equal.

  $10(.3 + .4) = 10 \times .7 = 7$

  $(10 \times .3) + (10 \times .4) = 3 + 4 = 7$

The distributive property of multiplication over subtraction:

- $3(5 - 2) = (3 \times 5) - (3 \times 2)$

  Simplifying the terms on each side of the equals sign shows that both sides are equal.

  $3(5 - 2) = 3(3) = 9$

  $(3 \times 5) - (3 \times 2) = 15 - 6 = 9$

- $\frac{1}{2}(\frac{1}{2} - \frac{1}{4}) = (\frac{1}{2} \times \frac{1}{2}) - (\frac{1}{2} \times \frac{1}{4})$

  Simplifying the terms on each side of the equals sign shows that both sides are equal.

  $\frac{1}{2}(\frac{1}{2} - \frac{1}{4}) = \frac{1}{2}(\frac{1}{4}) = \frac{1}{8}$

  $(\frac{1}{2} \times \frac{1}{2}) - (\frac{1}{2} \times \frac{1}{4}) = \frac{1}{4} - \frac{1}{8} = \frac{1}{8}$

- $.1(.5 - .4) = (.1 \times .5) - (.1 \times .4)$

  Simplifying the terms on each side of the equals sign shows that both sides are equal.

  $.1(.5 - .4) = .1(.1) = .01$

  $(.1 \times .5) - (.1 \times .4) = .05 - .04 = .01$

Inequality Symbols

$\neq$ is the symbol for "NOT EQUAL TO".

- $6 \neq 4$ 6 is not equal to 4

$<$ is the symbol for "LESS THAN".

- $2 < 9$ 2 is less than 9
- $\frac{1}{4} < \frac{1}{2}$ $\frac{1}{4}$ is less than $\frac{1}{2}$

$>$ is the symbol for "GREATER THAN".

- $10 > 6$ 10 is greater than 6
- $1 > .6$ 1 is greater than .6

$\leq$ is the symbol for "LESS THAN OR EQUAL TO".

- $18 \leq 20$ 18 is less than or equal to 20
- $18 \leq 18$ 18 is less than or equal to 18

$\geq$ is the symbol for "GREATER THAN OR EQUAL TO".

- $12 \geq 1$ 12 is greater than or equal to 1
- $12 \geq 12$ 12 is greater than or equal to 12

Tip: An inequality symbol always points to the smaller number. The "mouth" of the symbol wants to "eat" the larger number.

- $3 < 4$
- $4 > 3$
- $3 \leq 4$
- $4 \geq 3$

Inequality symbols can be used to describe a range of numbers.

- $0 < n < 5$ — Numbers between 0 and 5.
- $\frac{1}{4} < n < \frac{1}{2}$ — Numbers between $\frac{1}{4}$ and $\frac{1}{2}$.
- $7 > n > 5$ — Numbers between 5 and 7.
- $1 > n > .6$ — Numbers between .6 and 1.
- $18 \le n \le 26$ — Numbers between 18 and 26, including 18 and 26.
- $10 \ge n > 4$ — Numbers between 4 and 10, including 10, but excluding 4.

Graphing Numbers on a Number Line

The NUMBER LINE is a line on which every point corresponds to a number and every number to a point. It is divided into equal sections that represent a specific range of numbers.

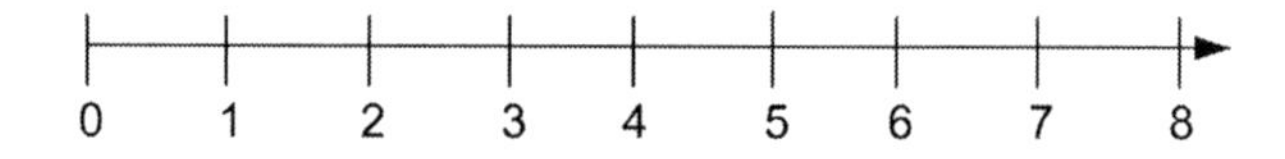

The number line allows us to visualize the order of numbers. A number to the right of any other number is the greater number.

Whatever numbers A and B represent, $A < B$ and $B > A$.

Individual numbers are graphed on the number line with dots.

- 4 is graphed on this number line.

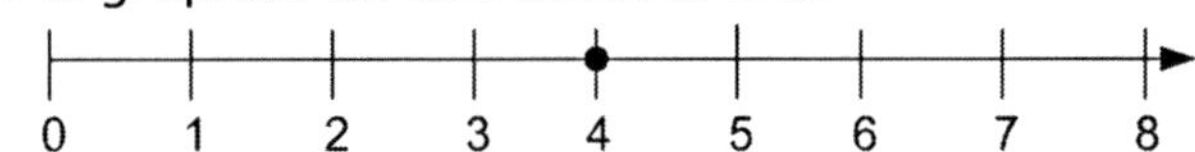

- 20 and 50 are graphed on this number line.

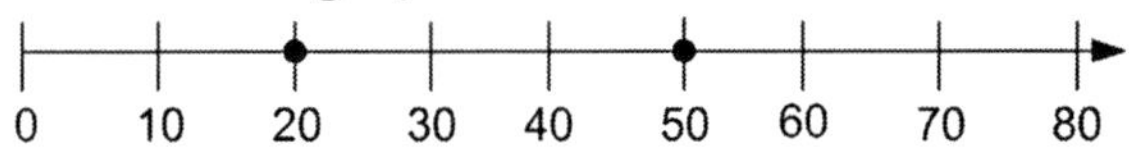

A set of continuous numbers is graphed on the number line as a thick line.

- All numbers greater than or equal to 3 and less than or equal to 6 ($3 \leq n \leq 6$) are graphed on this number line.

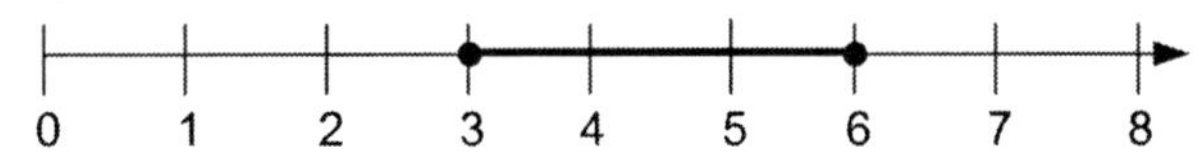

The numbers that are graphed include 3, 6, and all fractions and decimal numbers between them.

- All numbers greater than 3 and less than 6 ($3 < n < 6$) are graphed on this number line.

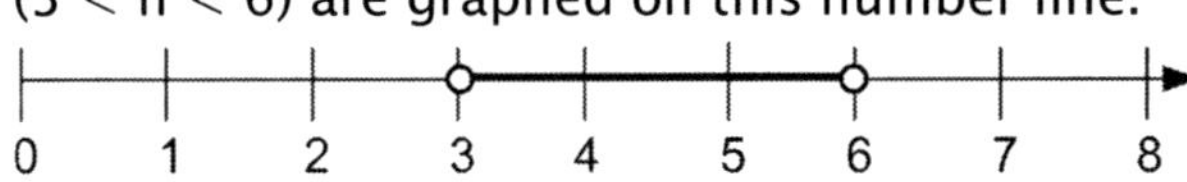

The numbers that are graphed do not include 3 or 6.

- All numbers greater than 3 and less than or equal to 6 ($3 < n \leq 6$) are graphed on this number line.

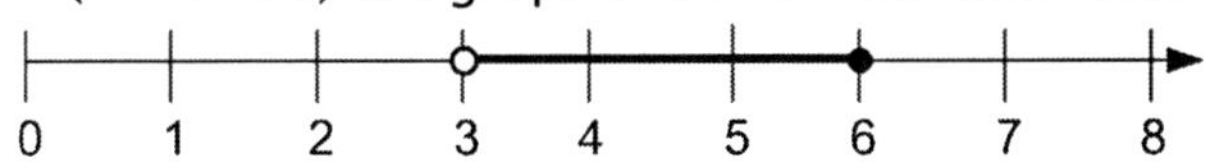

The numbers that are graphed include 6 but not 3.

Ratios, Proportions, and Percents

Ratio. Proportion. One hundred percent (%). One percent.

Ratio

A RATIO is a comparison of two quantities.

- If the lengths of two wooden boards are 8 ft. and 10 ft., the ratio of the shorter board to the longer board is 8 to 10.
  Besides saying "8 to 10", two other ways to write this ratio are: $\frac{8}{10}$ and 8:10.

A ratio is simplified the same way that a fraction is simplified.

- $\frac{8}{10} = \frac{4}{5}$
- 8 to 10 = 4 to 5
- 8:10 = 4:5

A ratio can be inverted and still be true.

- If the ratio of a short board to a longer board is 8 to 10, then the ratio of the long board to the short board is 10 to 8.
  Besides saying "10 to 8", two other ways to write this ratio are: $\frac{10}{8}$ and 10:8.

Proportion

Two equal ratios form a PROPORTION.

- $\frac{50 \text{ mi}}{4 \text{ gal}} = \frac{25 \text{ mi}}{2 \text{ gal}}$ is a proportion.

Tip: A different, but true proportion results when both fractions are replaced by their reciprocals.

- Since $\frac{50}{4} = \frac{25}{2}$, then $\frac{4}{50} = \frac{2}{25}$.

Tip: A different, but true proportion results when the numbers in opposite "corners" are swapped.

- Since $\frac{50}{4} = \frac{25}{2}$, then $\frac{2}{4} = \frac{25}{50}$ and $\frac{50}{25} = \frac{4}{2}$.

A proportion has equal cross-products.

- Since $\frac{50}{4} = \frac{25}{2}$, then $50 \times 2 = 4 \times 25$.

Percent

What percent means

ONE HUNDRED PERCENT (100%) of something represents all of it.

ONE PERCENT (1%) of something represents $\frac{1}{100}$ of it.

(1 percent literally means 1 per hundred.)

- 1% of 100 = 1
- 1% of 200 = 2
- 1% of 1 = $\frac{1}{100}$
- 1% of 2 = $\frac{2}{100}$
- 1% of a dollar is one cent.

Percents lets us compare different parts of a whole as if it were made up of 100 parts, no matter how many actual items make up the whole thing under consideration.

- If 25% of a collection of coins are dimes, then the ratio of dimes to all coins in the collection is 1 to 4.
  There could be 4 coins or 1,000 coins in the collection.

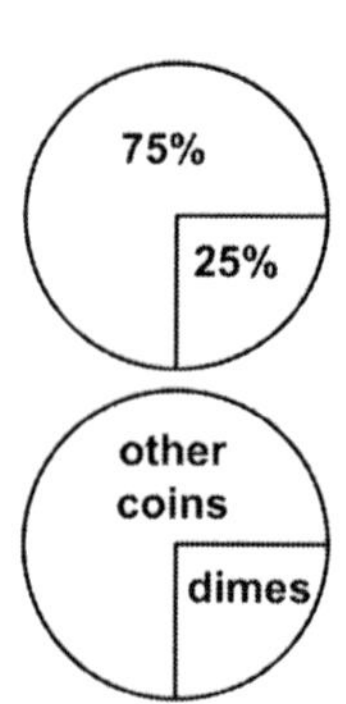

Percents and equivalent decimal numbers

Percents can be written as decimal numbers by moving the decimal point two places to the left.

- 50% = .5
- 100% = 1
- 25% = .25

- 10% = .1
- 3% = .03
- 2.5% = .025
- 150% = 1.5

Decimal numbers can be written as percents by moving the decimal point two places to the right.

- .25 = 25%
- 1 = 100%
- .62 = 62%
- .005 = .5%

Percents and equivalent fractions

To convert a percent to a fraction, place it over 100 and simplify the resulting fraction.

- $3\% = \frac{3}{100}$
- $50\% = \frac{50}{100} = \frac{1}{2}$

To convert a fraction to a percent, first convert the fraction to a decimal number. Then convert the decimal number to a percent.

- $\frac{1}{10} = .1 = 10\%$
- $\frac{1}{4} = .25 = 25\%$

Solving percent problems

Every percent problem has the following format:
PERCENT OF WHOLE = PART
where one of these three terms will be missing.

- Find 25% of 80. (Find the PART.)
- 25% of what number = 20? (Find the WHOLE.)
- What percent of 80 is 20? (Find the PERCENT.)

In a percent problem, "OF" means "TIMES".

Change the given format of a percent problem:
PERCENT of WHOLE = PART
into a format that is much more useful.

DECIMAL NUMBER X WHOLE = PART

- Find 25% of 80. → $.25 \times 80 = ?$
- 25% of what number = 20? → $.25 \times ? = 20$
- What percent of 80 is 20? → $? \times 80 = 20$

Tip: A useful strategy for recalling how to solve percent problems is to start with any simple multiplication fact. Then ask yourself how you would find the missing part.

- To find the answer to: $3 \times 4 = ?$, multiply 3×4.
- To find the answer to: $3 \times ? = 12$, divide $3\overline{)12}$.
- To find the answer to: $? \times 4 = 12$, divide $4\overline{)12}$.

To find the PART when you are given the percent and the whole, write the percent as a decimal number, and then multiply the decimal number by the whole.

- Problem: Find 25% of 80.

25% = .25

.25 × 80 = ?

$$\begin{array}{r} .25 \\ \times\ 80 \\ \hline 20.00 \end{array}$$

.25 × 80 = 20

25% of 80 is **20**.

(To find the answer to $3 \times 4 = ?$, multiply 3×4.)

To find the WHOLE when you are given the percent and the part, write the percent as a decimal number, then divide the part by the decimal number.

- Problem: 25% of what number = 20?

25% = .25

.25 × ? = 20

$$.25\overline{)20} = 25\overset{\displaystyle 80.}{\overline{)2000.}}$$

25% of **80** is 20.

(To find the answer to $3 \times ? = 12$, divide $3\overline{)12}$.)

To find the PERCENT when you are given the whole and the part, divide the part by the whole, and then write the decimal answer as a percent.

- Problem: What percent of 80 is 20?

  $? \times 80 = 20$

  $$80\overline{)20.00} = .25$$

  (To find the answer to $? \times 4 = 12$, divide $4\overline{)12}$.)

  Write the decimal as a percent by moving the decimal point two places to the right:
  $.25 = 25\%$.

  25% of 80 is 20.

NOTE: PROBLEMS INVOLVING THE PERCENT OF INCREASE OR DECREASE CAN BE SOLVED IN A SIMILAR WAY:

PERCENT X INITIAL VALUE = INCREASE OR DECREASE

If the value of something went from 80 to 100, by what percent did it increase?

Or,

If the value of something went from 80 to 60, by what percent did it decrease?

$? \times 80 = 20$ (increase or decrease)

There was a 25% (increase or decrease).

Percents greater than 100%

Solve problems that involve percents greater than 100% with the same techniques used to solve other percent problems.

- Find 110% of 80. Find the PART.

$1.10 \times 80 = ?$

$$\begin{array}{r} 1.1 \\ \times\ 80 \\ \hline 88.0 \end{array}$$

110% of 80 = **88**

- 110% of what number = 88? Find the WHOLE.

$1.1 \times ? = 88$

$$1.1\overline{)88} = 11.\overset{80.}{\overline{)880.}}$$

110% of **80** = 88

- What percent of 80 is 88? Find the PERCENT.

$? \times 80 = 88$

$$80\overline{)88} = 80\overset{1.1}{\overline{)88.0}}$$

$1.1 = 110\%$

110% of 80 = 88

Probability and Statistics (Selected Topics)

Probability. Data set. Average. Mean. Median. Mode. Range.

Probability of an Event

The PROBABILITY of an event occurring is the ratio of favorable outcomes to possible outcomes.

$$\text{Probability} = \frac{\text{number of Favorable Outcomes}}{\text{number of Possible Outcomes}}$$

- If you close your eyes and select one coin from a box containing 3 dimes, 1 nickel, and 6 pennies:

  The probability of selecting a dime is $\frac{3}{10}$.

  The probability of selecting a nickel is $\frac{1}{10}$.

  The probability of selecting a penny is $\frac{6}{10}$ or $\frac{3}{5}$.

The probability of several outcomes occurring is the product of each of their probabilities.

- The probability of tossing a coin three times and getting all heads is $\frac{1}{2} \times \frac{1}{2} \times \frac{1}{2} = \frac{1}{8}$.

HHH	THH
HHT	THT
HTH	TTH
HTT	TTT

Statistical Measures

Numbers collected for statistics are frequently referred to as scores, or DATA SETS.

The MEAN of a data set is its AVERAGE.

To find the mean of a data set, first add the scores, and then divide their sum by the number of scores.

- To find the mean of {12, 20, and 13}:
  First add them to get 45.
  Then divide 45 by 3 to get 15.

  15 is the mean (or average) of 12, 20, and 13.

The MODE of a data set is the most common score.

A data set can have no mode, one mode, or more than one mode.

- The set {12, 15, 20, 13, 15} has a mode of 15.
- The set {1, 2, 1, 2, 1} has a mode of 1.
- The set {6, 1, 2, 31, 2, 7, 6} has two modes, 2 and 6.
- The set {8, 6, 4, 5, 23, 2} has no mode.

The MEDIAN of a data set is the number in the middle of the scores after they are placed in order. (Half the scores are higher and half are lower.)

Tip: It does not matter whether the scores are written in ascending or descending order.

When there is an odd number of scores, the median is the middle score in the ordered list.

- To find the median of {12, 20, and 13}:

  First place them in order: 12, 13, 20.

  Then identify the middle score.

  13 is the median of 12, 20, and 13.

When there is an even number of scores, the median will be a number that is halfway between the two central scores in the ordered list.

Tip: When the halfway point is not obvious to you, find the average of the middle two scores.

- To find the median of {12, 20, 13, and 15}:

  First place them in order: 12, 13, 15, 20.

  Then determine the number that is halfway between the middle two scores.

  14 is the median of {12, 20, 13, 15}.

- The median of {4, 4, 5, 6} is 4.5.
- The median of {1, 2, 4, 4, 6, 8} is 4.

Tip: When you have to find both the median and the mean of a data set, it saves time to write the scores only once, beneath one another and in order. You can identify the median and then add the numbers to begin to find the mean.

The RANGE of a data set is the difference between the highest score and the lowest score.

- To find the range of {12, 20, 13, 15}:

  First find the highest and lowest scores. In this case, 12 and 20.

  Then subtract the lowest score from the highest.

  8 is the range of {12, 20, 13, 15}.

Geometry and Measurement
(Selected Topics)

Line segment. Angle. Degree. Right angle. Acute angle. Obtuse angle. Straight angle. Perpendicular lines. Parallel lines. Plane. Polygon. Regular polygon. Irregular polygon. Quadrilateral. Rectangle. Square. Parallelogram. Rhombus. Trapezoid. Triangle. Right triangle. Isosceles triangle. Equilateral triangle. Scalene triangle. Circle. Circumference. Radius. Diameter. Length. U.S. standard units of length. Inch (in). Foot (ft). Yard (yd). Mile (mi). Metric units of length. Millimeter (mm). Centimeter (cm). Meter (m). Kilometer (km). Perimeter. Area. Square units. Surface area. Rectangular prism. Volume. Cube.

Angles and Lines

Line segment

A line segment is a line connecting two points. The name of a line segment is written with a line over the names of its two end points.

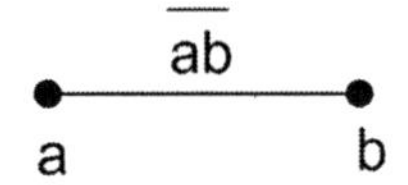

Angles

An ANGLE is formed when two straight lines intersect.

angle

A DEGREE is the unit of measurement of an angle.

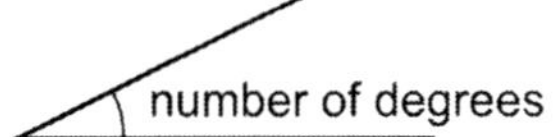

- "Zero degrees" is written as 0°.
- "Twenty-four degrees" is written as 24°.
- "One hundred degrees" is written as 100°.

A 360° angle is made by sweeping a line segment to form a circle.

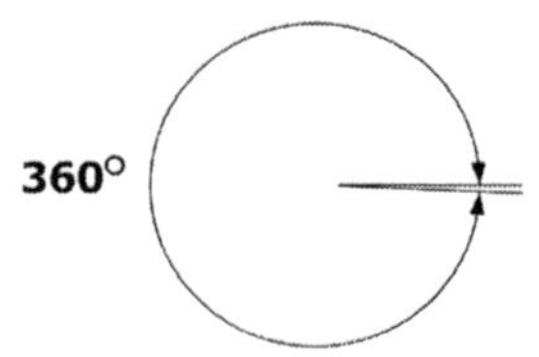

A 90° angle is called a RIGHT ANGLE.

A right angle is drawn with a small square at its center.

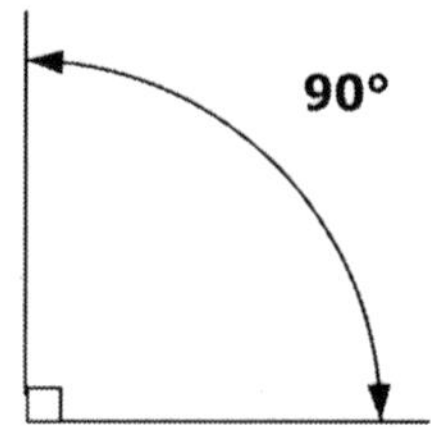

An angle between 0° and 90° is called an ACUTE ANGLE.

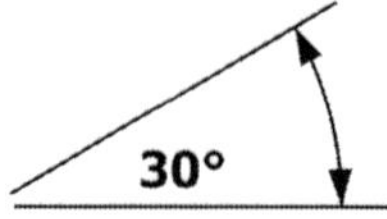

An angle greater than 90° is called an OBTUSE ANGLE.

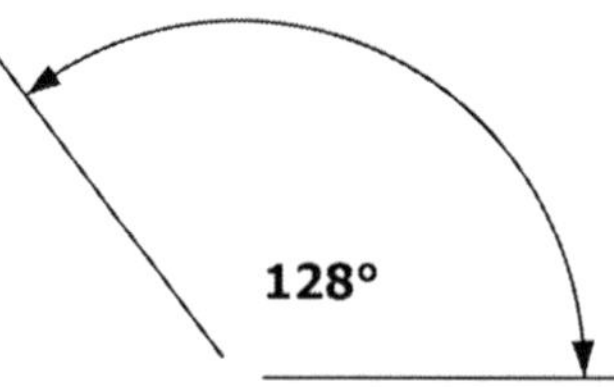

A 180° angle is called a STRAIGHT ANGLE.

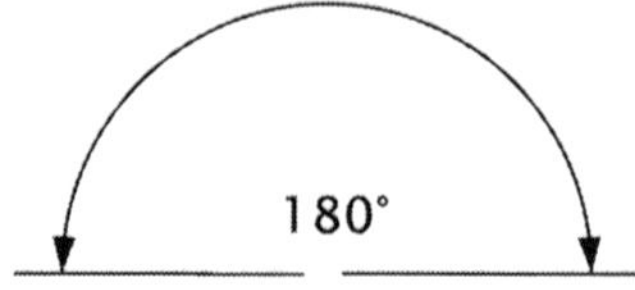

Perpendicular and parallel lines

PERPENDICULAR LINES are lines that intersect at a 90° angle.

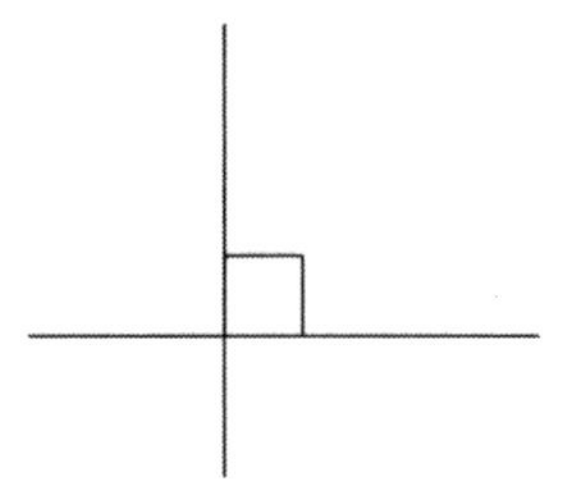

PARALLEL LINES are lines that run in the same direction and never intersect one another.

Polygons

Planes and polygons

In geometry, a flat surface is called a PLANE.

When lines enclose a flat space in a plane they create a figure called a POLYGON. (A many-sided plane figure.)

When the sides of a polygon are all equal to one another, the figure is called a REGULAR POLYGON.

When all the sides of a polygon are not equal to one another, the figure is called an IRREGULAR POLYGON.

Quadrilaterals

A QUADRILATERAL is a polygon that has four sides.

Rectangle

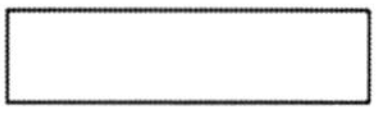

- Irregular polygon
- Opposite sides are equal and parallel
- Four 90° angles

Square

- Regular polygon; all four sides are equal
- Opposite sides are parallel
- Four 90° angles

Parallelogram

- Irregular polygon
- Opposite sides are equal and parallel
- Opposite angles are equal

Rhombus

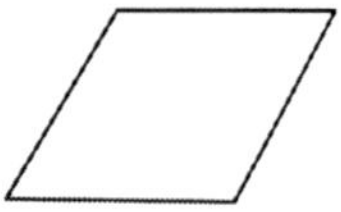

- Regular parallelogram; all four sides are equal
- Opposite sides are parallel
- Opposite angles are equal

Trapezoid

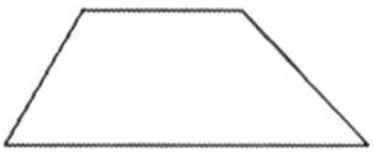

- Irregular polygon
- Two opposite sides are parallel

Triangles

A TRIANGLE is a polygon that has three sides and three angles that add up to 180°.

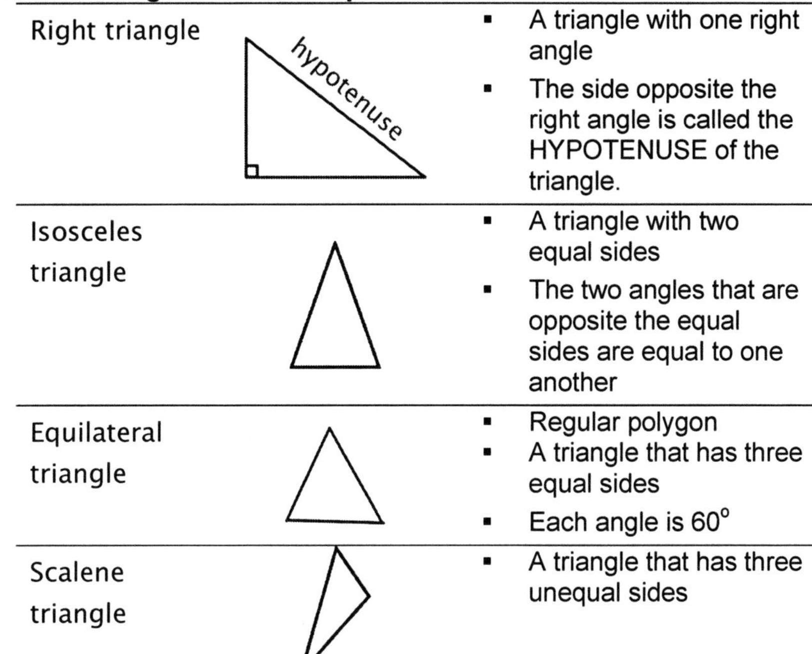

Right triangle	▪ A triangle with one right angle ▪ The side opposite the right angle is called the HYPOTENUSE of the triangle.
Isosceles triangle	▪ A triangle with two equal sides ▪ The two angles that are opposite the equal sides are equal to one another
Equilateral triangle	▪ Regular polygon ▪ A triangle that has three equal sides ▪ Each angle is 60°
Scalene triangle	▪ A triangle that has three unequal sides

The Pythagorean Theorem

The square of the hypotenuse of a right triangle equals the sum of the squares of the other two sides.

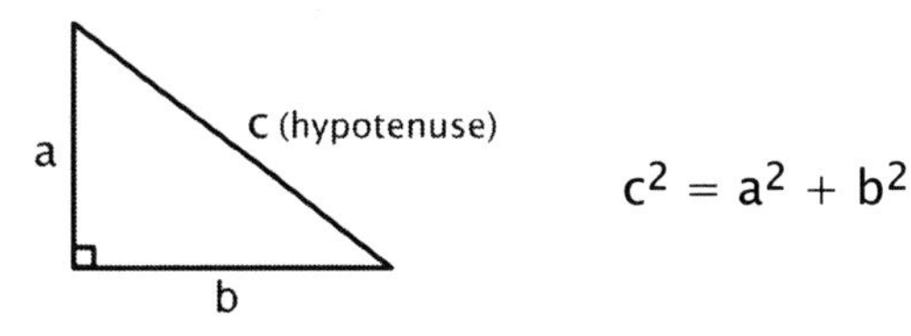

$$c^2 = a^2 + b^2$$

Frequently used examples of right triangles include:

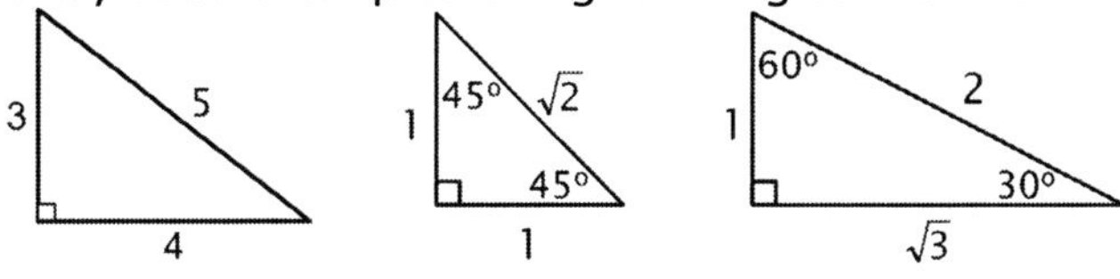

Circles

The CIRCUMFERENCE of a circle is the length of the outside edge of the circle.

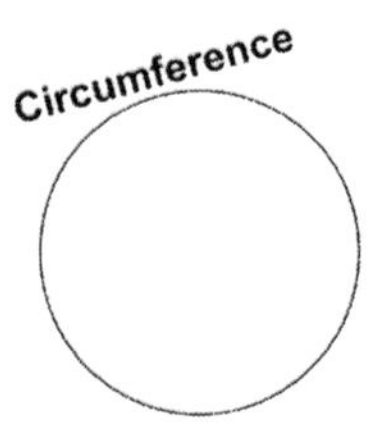

The RADIUS of a circle is the distance from its center to its outside edge.

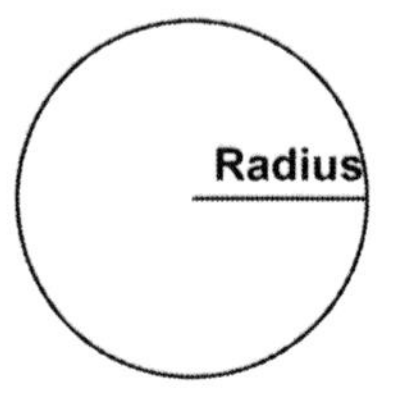

The DIAMETER of a circle is the length of a line that cuts the circle in half. The diameter of a circle passes through the center of the circle, and is twice as long as the radius.

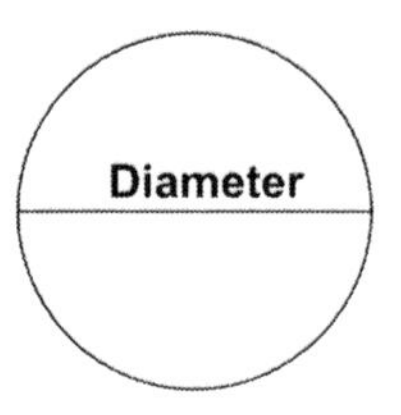

The parts of any circle have fixed ratios.

$$\frac{\text{diameter}}{\text{radius}} = 2$$

$$\text{diameter} = 2 \times \text{radius}$$

The diameter of a circle is twice as long as the radius.

$$\frac{\text{circumference}}{\text{diameter}} = \pi$$

$$\text{circumference} = \pi \times \text{diameter}$$

$$\text{circumference} = 2 \times \pi \times \text{radius}$$

The ratio of the circumference to the diameter is represented by the Greek letter π ("**pi**").

π is a non-terminating, non-repeating decimal number.

$$\pi = 3.14159265358979323846264338327950...$$

π is approximated in whole number and decimal calculations as 3.14 or 3.1416.

- The circumference of a circle that has a diameter of 5 inches is approximately $3.14 \times 5 = 15.7$ inches.

π is approximated in fractional calculations as $3\frac{1}{7}$ or $\frac{22}{7}$.

- The circumference of a circle that has a diameter of $\frac{1}{2}$ an inch is approximately

  $\frac{22}{7} \times \frac{1}{2} = \frac{22}{14} = 1\frac{8}{14} = 1\frac{4}{7}$ inches.

Length

LENGTH is a measurement of distance.

U.S. STANDARD UNITS OF LENGTH include: inches (in), feet (ft), yards (yd), and miles (mi).

1 foot (ft) = 12 inches (in)

1 yard (yd) = 3 feet = 36 inches

1 mile (mi) = 5280 feet

A DOUBLE QUOTATION MARK is sometimes used to indicate inches.

- $3'' = 3$ inches
- On a ruler, $1''$ is usually divided into $\frac{1}{16}''$, $\frac{1}{8}''$, $\frac{1}{4}''$, and $\frac{1}{2}''$ spaces.

A SINGLE QUOTATION MARK is sometimes used to indicate feet.

- $5' = 5$ feet
- 1 yard (yd) $= 3' = 36''$

METRIC UNITS OF LENGTH include: millimeters (mm), centimeters (cm), meters (m), and kilometers (km).

1 kilometer (km)	= 1000 meters
1 meter (m)	= 100 centimeters = 1000 millimeters
1 centimeter (cm)	= 10 millimeters = .01 meters
1 millimeter (mm)	= .001 meters

Tip: An inch is about as long as two and a half centimeters.

$1'' = 2.54$ cm

Tip: A meter is about 3 inches longer than a yard.

$1 \text{ m} = 39.37''$

Perimeter

The PERIMETER of a polygon is the sum of the lengths of all of its sides.

The perimeter of a rectangle is twice the sum of the length plus width.

$2 \times (\text{LENGTH} + \text{WIDTH})$

The perimeter of a square is four times the length of one side.

$4 \times \text{SIDE}$

The perimeter of a rhombus is four times the length of one side.

$4 \times \text{SIDE}$

The perimeter of an equilateral triangle is three times the length of one side.

$3 \times \text{SIDE}$

Tip: A common problem involves finding the PERIMETER OF A RIGHT TRIANGLE when only two sides are given.
Use the Pythagorean Theorem to find the missing side, and then add the three sides together.

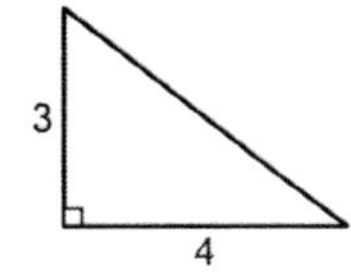

The hypotenuse of this triangle equals the square root of $3^2 + 4^2$.
Since $\sqrt{9+16} = 5$, the perimeter of the triangle is $3 + 4 + 5 = 12$.

Area

AREA is a measurement of a surface, and is measured in SQUARE UNITS.

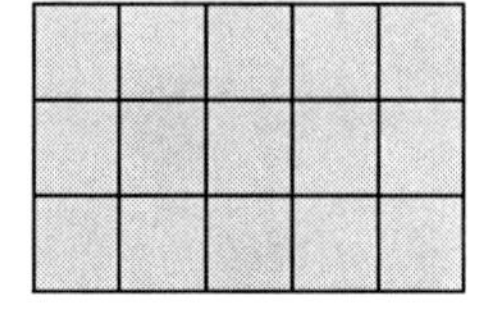

This surface has an area of 15 square units.

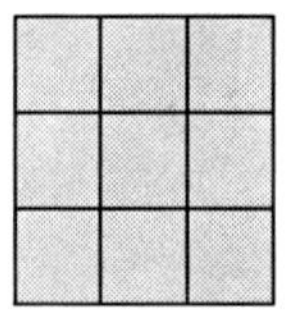

This surface has an area of 9 square units.

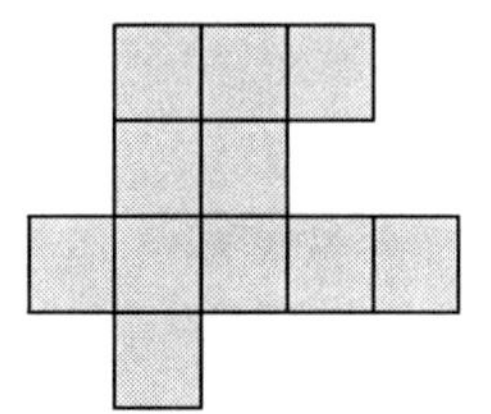

This surface has an area of 11 square units.

U.S. Standard units of area include: square inches (in^2), square feet (ft^2), square yards (yd^2), and square miles (mi^2).

- If the units in the previous examples were inches, then the three figures would have the following areas: 15 in^2, 9 in^2, and 11 sq. in.

Metric units of area include: square centimeters (cm^2), square meters (m^2), and square kilometers (km^2).

The area of a rectangle , parallelogram, or rhombus is base times height.

16 2
8

8 x 2 = 16 square units

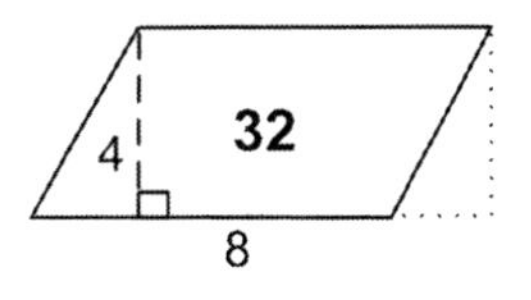

8 x 4 = 32 square units

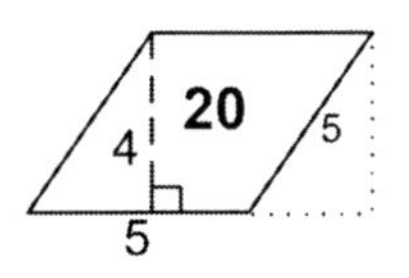

5 x 4 = 20 square units

The area of a square is the square of a side.

9 3

$3^2 = 9$ square units

The area of a triangle is half the base times height.

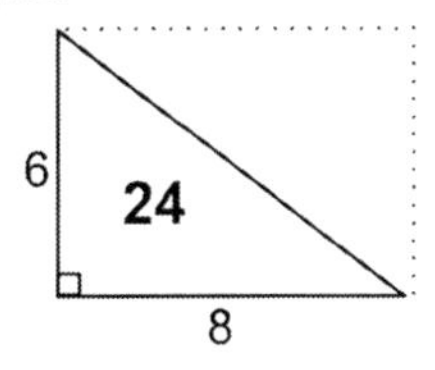

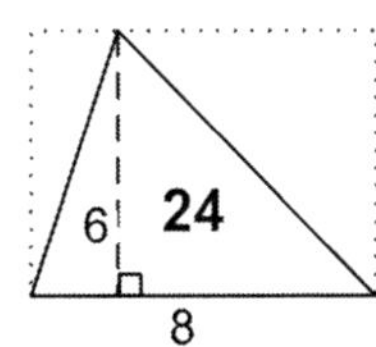

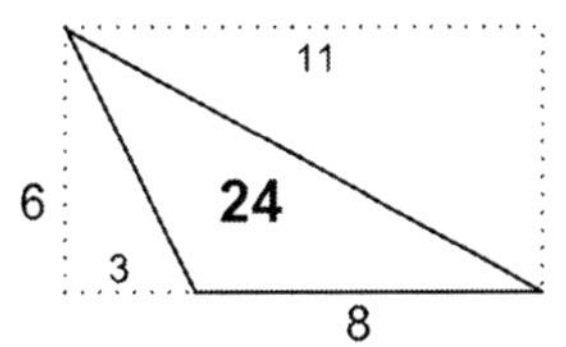

$\frac{6 \times 8}{2} = 24$ square units

The area of a circle is π times the radius squared.

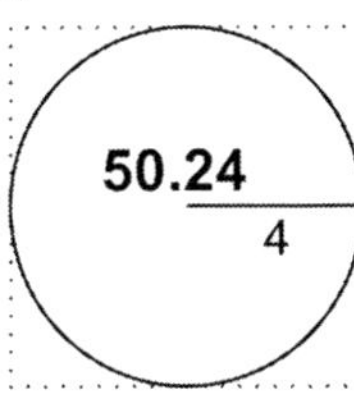

$3.14 \times 4^2 = 50.24$ square units

Surface area

SURFACE AREA is a measure of the total area of every surface of a solid object.

A brick-shaped object is called a RECTANGULAR PRISM. The surface area of a rectangular prism is the sum of the areas of all six surfaces: top, bottom, front, back, left side, and right side.

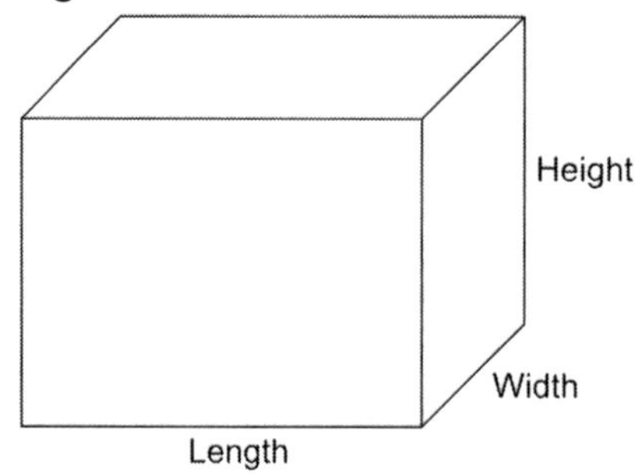

Top = length × width
Bottom = length × width

Front = length × height
Back = length × height

Right side = width × height
Left side = width × height

Tip: To quickly find the surface area of a rectangular prism, find the area of the three visible surfaces; then double the result.

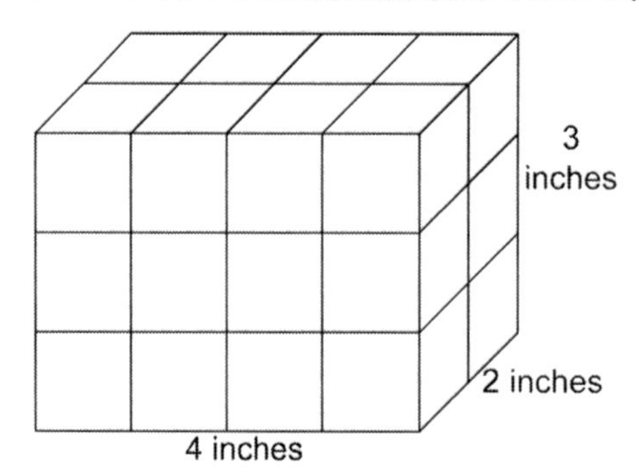

Top: $4 \times 2 = 8$ in^2

Front: $4 \times 3 = 12$ in^2

Right: $2 \times 3 = 6$ in^2

$8 + 12 + 6 = 26$ in^2

$26 \times 2 = 52$ in^2

The surface area of this prism is 52 square inches.

Volume

**VOLUME is a measure of occupied space.
Volume is measured in cubic units.**

A CUBE is a rectangular prism having edges of equal length. The volume of a cube is length × width × height (or edge × edge × edge).

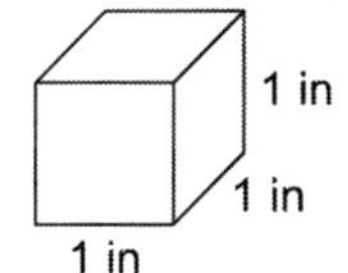

The volume of a cube having a 1 inch edge is $1 \times 1 \times 1 = 1$ cubic inch.

The volume of a RECTANGULAR PRISM is length × width × height.

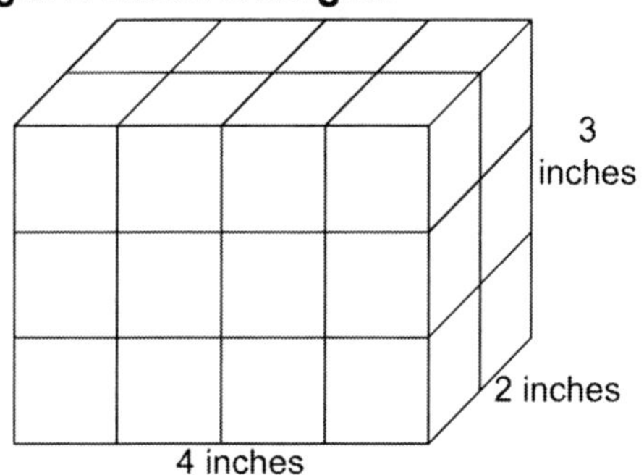

This rectangular prism has a volume of $4 \times 2 \times 3 = 24$ in^3.

Positive and Negative Numbers

Positive number. Negative number. Positive sign. Negative sign. Number line. Signed number. Absolute value | |. Opposite. Integer. Rational number. Irrational number. Real number. Negative base. Negative exponent.

Signed Numbers and the Number Line

Numbers that are greater than a reference point are POSITIVE. Numbers that are less than a reference point are NEGATIVE.

- If the reference point is having no money, having money is positive and owing money is negative.
- Having $2 can be described as positive 2 dollars.
- Owing $3 can be described as negative 3 dollars.

- If the reference point on a thermometer is zero degrees, temperatures above zero are positive and temperatures below zero are negative.
- 30° can be described as positive 30 degrees.
- Two degrees below zero can be described as negative 2 degrees.

- If the reference point is sea level, the altitudes of places on earth that are above sea level are positive and altitudes below sea level are negative.
- 500 feet above sea level can be described as positive 500 feet.
- 20 feet below sea level can be described as negative 20 feet.

Positive numbers are written with a POSITIVE SIGN ($^{+}$).

Negative numbers are written with a NEGATIVE SIGN ($^{-}$).

- Having \$2 can be represented as $^{+}2$. ("positive 2")
- Owing \$3 can be represented as $^{-}3$. ("negative 3")
- 30° can be represented as $^{+}30$.
- Two degrees below zero can be represented as $^{-}2$.
- 500 feet above sea level can be represented as $^{+}500$.
- 20 feet below sea level can be represented as $^{-}20$.

Tip: Positive and negative signs are usually written as raised symbols, but need not be.

- $^{-}3$ can be written as -3.
- $^{+}5$ can be written as +5.

NOTE: THESE ARE NOT MINUS AND PLUS SIGNS BECAUSE THERE ARE NO NUMBERS IN FRONT OF THEM.

Positive and negative numbers can be shown on a NUMBER LINE that has 0 as a reference point.

0 is called the ORIGIN.

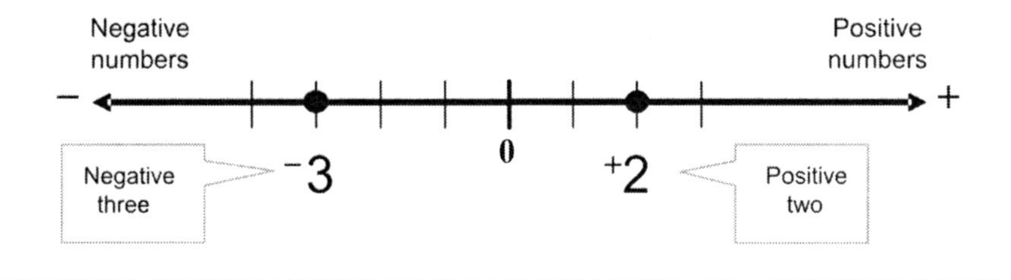

Sign and Absolute Value

A positive or negative number is called a SIGNED NUMBER because it either has a positive sign or a negative sign.

The sign tells us whether the number is greater than zero (positive) or less than zero (negative).

Zero has no sign because it is the reference point.

Any non-zero number that has no sign is positive.

- $2 = {}^{+}2$
- $10 = {}^{+}10$

The numeric part of a signed number tells us how far from 0 the number is.

- ${}^{-}5$ and ${}^{+}5$ have both a sign and a numeric term of 5.

  ${}^{-}5$ is five units to the left of 0.

  ${}^{+}5$ is five units to the right of 0.

The numeric term in a signed number is called its ABSOLUTE VALUE.

- The value of ${}^{+}5$ is positive five.
  Its *absolute* value is five.
- The value of ${}^{-}5$ is negative five.
  Its *absolute* value is five.

The mathematical symbol for the absolute value of a number is two straight lines that sandwich the number between them.

- $|^{-}7|$ = The absolute value of $^{-}7 = 7$
- $|^{+}10|$ = The absolute value of $^{+}10 = 10$
- $|^{-}5| = 5$
- $|^{+}5| = 5$

NOTE: IF A NUMBER HAS NO SIGN, IT IS POSITIVE.

- $|5| = |^{+}5| = 5$
- $|12| = |^{+}12| = 12$

The absolute value of any number is always positive because it is the distance of that number from zero, regardless of whether it is to the right or left of zero.

Tip: To write the absolute value of any number, just discard its sign.

|0| = 0 because the distance of zero from zero (the reference point) is zero.

Opposites

The OPPOSITE of a signed number is a number with the same absolute value, but a different sign. That is, the number on the opposite side of zero that is the same distance from zero.

- $^{-}5$ and $^{+}5$ are opposites of one another.

  The opposite of $^{-}5$ is $^{+}5$.

  The opposite of $^{+}5$ is $^{-}5$.

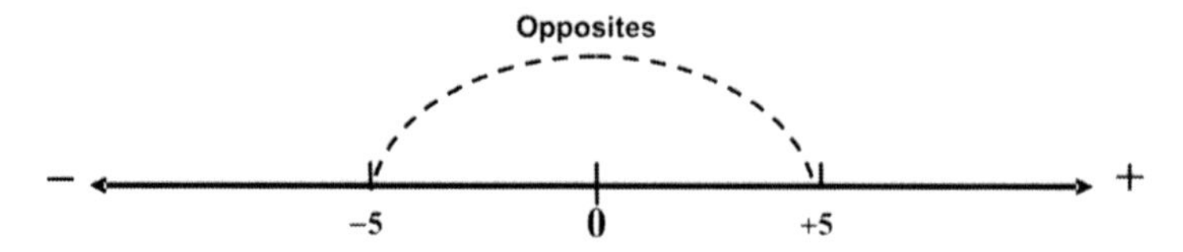

The mathematical symbol for opposite is a dash.

- The opposite of $^{-}5$ is written as $-(^{-}5)$.
- The opposite of $^{+}5$ is written as $-(^{+}5)$.
- The opposite of 5 is written as -5, which also can be read as negative five.

The opposite of 0 is 0.

- $-0 = 0$

Tip: A dash is used to represent three different mathematical concepts: subtraction, a negative sign, and the opposite of a signed number. A number must precede a dash when it means "minus".

When there are several adjacent signs in front of an absolute value, they can all be replaced by a single positive or negative sign.

Steps

1. Count the dashes (opposite symbols and negative signs) that precede the absolute value of a number.
2. Then ...
   - IF THERE IS AN EVEN NUMBER OF DASHES, replace all symbols with a single positive sign.
     - $-\ [-\ (^{+}3)] = {}^{+}3$

       This is read as the opposite of the opposite of positive three.

       The opposite of $^{+}3$ is $^{-}3$, and the opposite of $^{-}3$ is $^{+}3$.
     - $-\{-\ [-\ (^{-}3)]\} = {}^{+}3$
   - IF THERE IS AN ODD NUMBER OF DASHES, replace all symbols with a single negative sign.
     - $-\ [-\ (^{-}3)] = {}^{-}3$
     - $-\{-\ [-\ (^{+}3)]\} = {}^{-}3$

Integers, Rational and Irrational Numbers, and Real Numbers

Signed numbers are classified as integers, rational numbers, irrational numbers, and real numbers.

The INTEGERS are all the whole numbers and their opposites.

- Integers include:

  $^{-}5,\ ^{-}4,\ ^{-}3,\ ^{-}2,\ ^{-}1,\ 0,\ ^{+}1,\ ^{+}2,\ ^{+}3,\ ^{+}4,\ ^{+}5$

RATIONAL NUMBERS are numbers that can be written as fractions (ratios). They include all the integers, positive and negative fractions, terminating decimal numbers, and non-terminating repeating decimal numbers.

- Rational numbers include:

  $^{-}5,\ ^{-}2\frac{1}{4},\ ^{-}.\overline{3},\ -\frac{1}{4},\ 0,\ ^{+}.405,\ ^{+}1,\ ^{+}2.7$

IRRATIONAL NUMBERS are numbers that cannot be written as fractions. They include all non-terminating, non-repeating decimal numbers.

- Irrational numbers include:

$$\sqrt{2} = 1.4142135623730950488016887242097...$$

$$\sqrt{3} = 1.7320508075688772935274463415059...$$

$$\sqrt{5} = 2.2360679774997896964091736687313...$$

$$\sqrt{6} = 2.4494897427831780981972840747059...$$

$$\sqrt{7} = 2.6457513110645905905016157536393...$$

$$\sqrt{8} = 2.8284271247461900976033774484194...$$

$$\pi = 3.1415926535897932384626433832795...$$

The REAL numbers include all rational and irrational numbers.

- Real numbers include:

$$^{-}5,\ ^{-}2\frac{1}{4},\ ^{-}\sqrt{5},\ ^{-}.\bar{3},\ ^{-}\frac{1}{4},\ 0,\ ^{+}.405,\ ^{+}1,\ ^{+}2.7,\ \pi$$

Ordering Positive and Negative Numbers

Tip: The best way to order signed numbers is to visualize them on a number line.

On a number line, just as with whole numbers, a number to the left of another number is the lesser number.

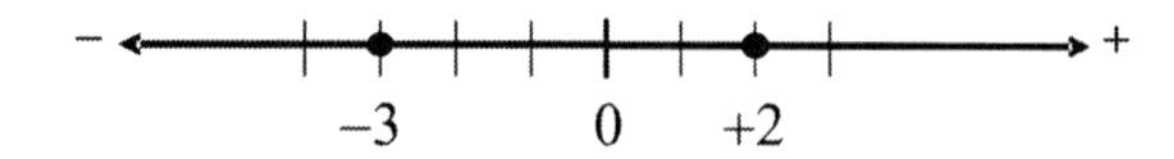

- $^-3 < 0$ Someone who owes $3 has less money than someone who has no money.

  A fish 3 feet below sea level is lower than a leaf floating on the water.

  Negative 3 degrees is colder than 0 degrees.

- $^-3 < {}^+2$

- $0 < {}^+2$

- $^-3$ is less than both 0 and 2 even though it has the greater absolute value.

- $^-1{,}000{,}000 < {}^-5$ because $^-1{,}000{,}000$ is much more negative.

Ordering rational numbers that are written as fractions or decimal numbers is the same as ordering integers: a number that is more negative is the lesser one.

- $^{-}\frac{1}{2} < ^{-}\frac{1}{4}$
- $^{-}5 < ^{-}\frac{1}{2}$
- $^{-}11 < ^{+}\frac{1}{2}$
- $^{-}3.5 < ^{-}1.9$
- $^{-}.2 < ^{-}.02$

Adding Two Signed Numbers

When two signed numbers that are being added have the same sign, add their absolute values. Give the answer the same sign as the numbers being added.

- ${}^{+}7 + {}^{+}3 = {}^{+}10$ — Since $+7$ and $+3$ are both positive numbers, add their absolute values. The answer is positive.
- ${}^{-}7 + {}^{-}3 = {}^{-}10$ — Since ${}^{-}7$ and ${}^{-}3$ are both negative numbers, add their absolute values. The answer is negative.

When two signed numbers that are being added have different signs, subtract the smaller absolute value from the larger one. Give the answer the same sign as the number with the larger absolute value.

- ${}^{+}7 + {}^{-}3 = {}^{+}4$
- ${}^{-}3 + {}^{+}7 = {}^{+}4$

Since ${}^{+}7$ and ${}^{-}3$ have different signs, subtract their absolute values. Since 7 is greater than 3, the answer is given the sign of the 7, positive.

- ${}^{-}7 + {}^{+}3 = {}^{-}4$
- ${}^{+}3 + {}^{-}7 = {}^{-}4$

Since ${}^{-}7$ and ${}^{+}3$ have different signs, subtract their absolute values. Since 7 is greater than 3, the answer is given the sign of the 7, negative.

Adding a signed number to its opposite results in 0.

- ${}^{+}5 + {}^{-}5 = 0$

Use the same procedure when numbers have no signs. Treat unsigned numbers as positive numbers.

- $7 + 3 = {}^{+}10$ The signs are the same.
- $7 + {}^{-}3 = {}^{+}4$ The signs are different and $7 > 3$.
- ${}^{-}7 + 3 = {}^{-}4$ The signs are different and $7 > 3$.

Use the same procedure to add signed fractions or decimal numbers.

- $+\frac{3}{5} + +\frac{1}{5} = +\frac{4}{5}$
- $-\frac{3}{5} + +\frac{1}{5} = -\frac{2}{5}$
- ${}^{+}1.5 + {}^{-}.3 = {}^{+}1.2$
- ${}^{-}1.5 + {}^{-}.3 = {}^{-}1.8$

Subtracting One Signed Number From Another

Instead of subtracting a number, add its opposite.

Adding the opposite of a number gives the same result as subtracting it.

- $7 - 3 = 4$.

  $7 + {}^{-}3$ also equals 4.

Convert every subtraction problem into one of addition, then use the same procedure used to add signed numbers.

Steps

1. Replace the minus sign with a plus sign.
2. Change the sign of the number being subtracted from positive to negative or negative to positive. (This replaces the number being subtracted with its opposite.)
3. Add the two numbers, using the steps for adding signed numbers.

- ${}^{+}7 - {}^{+}3 = {}^{+}7 + {}^{-}3 = {}^{+}4$
- ${}^{-}7 - {}^{-}3 = {}^{-}7 + {}^{+}3 = {}^{-}4$
- ${}^{+}7 - {}^{-}3 = {}^{+}7 + {}^{+}3 = {}^{+}10$
- ${}^{-}7 - {}^{+}3 = {}^{-}7 + {}^{-}3 = {}^{-}10$

- ${}^{+}3 - {}^{+}7 = {}^{+}3 + {}^{-}7 = {}^{-}4$
- ${}^{-}3 - {}^{-}7 = {}^{-}3 + {}^{+}7 = {}^{+}4$
- ${}^{-}3 - {}^{+}7 = {}^{-}3 + {}^{-}7 = {}^{-}10$
- ${}^{+}3 - {}^{-}7 = {}^{+}3 + {}^{+}7 = {}^{+}10$

Use the same procedure when numbers have no signs. Treat unsigned numbers as positive numbers.

- $7 - 3 = 7 + {}^{-}3 = {}^{+}4$
- $7 - {}^{-}3 = 7 + {}^{+}3 = {}^{+}10$
- ${}^{-}7 - 3 = {}^{-}7 + {}^{-}3 = {}^{-}10$

Use the same procedure to subtract signed fractions or decimal numbers.

- $\frac{-3}{5} - \frac{+1}{5} = \frac{-3}{5} + \frac{-1}{5} = \frac{-4}{5}$
- $\frac{-3}{5} - \frac{-1}{5} = \frac{-3}{5} + \frac{+1}{5} = \frac{-2}{5}$

- ${}^{+}1.5 - {}^{+}.3 = {}^{+}1.5 + {}^{-}.3 = {}^{+}1.2$
- ${}^{+}1.5 - {}^{-}.3 = {}^{+}1.5 + {}^{+}.3 = {}^{+}1.8$

Expert Strategies for Adding and Subtracting Signed Numbers

Treat all signs: plus, minus, positive, and negative signs as symbols.

Steps

1. Replace groups of adjacent symbols with a single symbol by counting the "–" symbols in the group.
   - IF THERE IS AN EVEN NUMBER OF "–" SYMBOLS IN THE GROUP, replace them all with a single "+" symbol.
   - IF THERE IS AN ODD NUMBER OF "–" SYMBOLS IN THE GROUP, replace them all with a single "–" symbol.
2. Note the symbols that are in front of each absolute value.
   - IF THE SYMBOLS ARE THE SAME, add the absolute values and give the answer the same symbol (sign).
   - IF THE SYMBOLS ARE DIFFERENT, subtract the smaller absolute value from the larger one, and give the answer the same symbol (sign) as the number with the larger absolute value.
     - $^{+}7 + {}^{+}3 = 7 + 3 = 10$
     - $^{-}7 + {}^{-}3 = -7 - 3 = -10$
     - $^{+}7 - {}^{+}3 = 7 - 3 = 4$
     - $^{-}7 - {}^{-}3 = -7 + 3 = -4$
     - $^{+}7 - {}^{-}3 = 7 + 3 = 10$
     - $^{-}7 - {}^{+}3 = -7 - 3 = -10$
     - $^{+}7 + {}^{-}3 = 7 - 3 = 4$
     - $^{-}7 + {}^{+}3 = -7 + 3 = -4$

Tip: Ignore the fact that addition problems can become subtraction problems, and vice-versa. This technique gives the correct answer and is extremely useful when solving algebra problems.

The technique works for all symbols that are grouped together, including opposite signs.

- $----^{-}3 = {}^{-}3$
- $---^{-}3 = {}^{+}3$
- $----^{+}9 = {}^{+}9$
- $-^{+}9 = {}^{-}9$
- $-(+7) + [-(+3)] = -7 - 3 = -10$
- $-7 + [-(-3)] = -7 + 3 = -4$

Tip: When adding or subtracting several signed numbers, place the numbers into two groups: plus and minus. Then subtract the minus group from the plus group.

<u>Steps</u>

1. Replace groups of symbols with a single + or – symbol.
2. Add the absolute values that are preceded by + symbols.
3. Add the absolute values that are preceded by – symbols.
4. Subtract the sum in Step 3 from the sum in Step 2.

- $+1 + ({}^{-}4) - ({}^{-}2) - 7 + ({}^{-}3)$

  $= +1 - 4 + 2 - 7 - 3$

  $+1 \boxed{-4} + 2 \boxed{-7} \boxed{-3}$

  $= +1 + 2 \boxed{-4 - 7 - 3}$

  $= (1 + 2) - (4 + 7 + 3)$

  $= 3 - 14$

  $= -11$

Multiplying Signed Numbers

When two signed numbers that are being multiplied have the same sign, multiply their absolute values and make the result positive.

- $^{+}5 \times (^{+}6) = {}^{+}30$ $\quad$ $^{+}6 + {}^{+}6 + {}^{+}6 + {}^{+}6 + {}^{+}6 = {}^{+}30$
- $^{-}5(^{-}6) = {}^{+}30$ $\quad$ (In this case, multiplication is not repeated addition.)

When two signed numbers to be multiplied have different signs, multiply their absolute values and make the result negative.

- $(^{-}5)(^{+}6) = {}^{-}30$ $\quad$ $^{-}5 + {}^{-}5 + {}^{-}5 + {}^{-}5 + {}^{-}5 + {}^{-}5 = {}^{-}30$
- $^{+}5(^{-}6) = {}^{-}30$ $\quad$ $^{-}6 + {}^{-}6 + {}^{-}6 + {}^{-}6 + {}^{-}6 = {}^{-}30$

The product of two negative numbers must be positive because the distributive property must hold true for all numbers.

An example demonstrates this:

$0 = {}^{-}5(0)$ $\quad$ Any number times 0 equals 0

$0 = {}^{-}5(^{+}6 + {}^{-}6)$ $\quad$ Substitute $(^{+}6 + {}^{-}6)$ for 0

$0 = {}^{-}5(^{+}6) + {}^{-}5(^{-}6)$ $\quad$ Use the distributive property

$0 = {}^{-}30 + {}^{-}5(^{-}6)$ $\quad$ $^{-}5(^{+}6) = {}^{-}30$

Since $0 = {}^{-}30 + {}^{+}30$, $^{-}5(^{-}6)$ must equal $^{+}30$.

Use the same procedure when numbers have no signs. Treat unsigned numbers as positive numbers.

- $5 \times (^{+}6) = {}^{+}30$
- $(^{-}5)(6) = {}^{-}30$
- $5(^{-}6) = {}^{-}30$

Multiplying any number by –1 results in the opposite of the number.

- $(^{-}1)(^{+}10) = {}^{-}10$
- $^{-}1 \cdot (^{-}10) = {}^{+}10$
- $^{+}10 \times (^{-}1) = {}^{-}10$
- $^{-}10\,(^{-}1) = {}^{+}10$

When more than two signed numbers are being multiplied, the sign of the answer depends on how many negative numbers are being multiplied.

Steps

1. Write the product of all the absolute values.
2. Count the number of negative numbers that are being multiplied.
3. Give the result a sign:

- IF THERE IS AN EVEN NUMBER OF NEGATIVE NUMBERS, make the result positive.
  - $^{-}3(^{+}2)(^{-}10) = {}^{+}60$
  - $2(^{+}1)(^{+}3)(^{+}3) = {}^{+}18$
- IF THERE IS AN ODD NUMBER OF NEGATIVE NUMBERS, make the result negative.
  - $^{-}5(^{+}4)(^{+}2) = {}^{-}40$
  - $^{+}3(^{-}1)(^{-}2)(^{-}2) = {}^{-}12$

Use the same procedure to multiply signed fractions or decimal numbers.

- $^{-}\frac{1}{2} \times {}^{+}\frac{3}{4} = {}^{-}\frac{3}{8}$
- $^{-}\frac{1}{2} \times {}^{-}\frac{3}{4} = {}^{+}\frac{3}{8}$
- $^{-}\frac{1}{2} \times {}^{-}\frac{1}{2} \times {}^{-}\frac{3}{4} = {}^{-}\frac{3}{16}$
- $^{-}1 \times {}^{+}\frac{1}{2} = {}^{-}\frac{1}{2}$
- $^{+}1.5 \times {}^{+}.3 = {}^{+}.45$
- $^{-}1.5 \times {}^{-}.3 = {}^{+}.45$
- $^{-}2 \times {}^{+}.2 \times {}^{-}.3 = {}^{+}.12$
- $^{-}1 \times {}^{-}.5 = {}^{+}.5$

Dividing One Signed Number by Another

NOTE: THE METHOD USED TO FIND THE SIGN OF THE RESULT OF A DIVISION PROBLEM IS SIMILAR TO THAT USED WHEN MULTIPLYING TWO SIGNED NUMBERS.

When the numbers to be divided have the same sign, divide their absolute values and make the result positive.

- $^{+}8 \div {}^{+}2 = {}^{+}4$ CHECK: $^{+}2 \times {}^{+}4 = {}^{+}8$
- $^{-}8 \div {}^{-}2 = {}^{+}4$ CHECK: $^{-}2 \times {}^{+}4 = {}^{-}8$

When the numbers to be divided have different signs, divide their absolute values and make the result negative.

- $^{-}8 \div {}^{+}2 = {}^{-}4$ CHECK: $^{+}2 \times {}^{-}4 = {}^{-}8$
- $^{+}8 \div {}^{-}2 = {}^{-}4$ CHECK: $^{-}2 \times {}^{-}4 = {}^{+}8$

Treat unsigned numbers as positive numbers.

- $8 \div {}^{-}2 = {}^{-}4$ CHECK: $^{-}2 \times {}^{-}4 = 8$
- $^{-}8 \div 2 = {}^{-}4$ CHECK: $2 \times {}^{-}4 = {}^{-}8$

Dividing any number by –1 results in the opposite of the number.

- $^{+}8 \div {}^{-}1 = {}^{-}8$ CHECK: $^{-}1 \times {}^{-}8 = {}^{+}8$
- $^{-}8 \div {}^{-}1 = {}^{+}8$ CHECK: $^{-}1 \times {}^{+}8 = {}^{-}8$

Dividing any number by its opposite results in –1.

- $^{+}8 \div {}^{-}8 = {}^{-}1$ CHECK: $^{-}8 \times {}^{-}1 = {}^{+}8$
- $^{-}8 \div {}^{+}8 = {}^{-}1$ CHECK: $^{+}8 \times {}^{-}1 = {}^{-}8$

Use the same procedure to divide signed fractions or decimal numbers.

- $^{-}\frac{1}{2} \div {}^{-}\frac{3}{4} = {}^{-}\frac{1}{2} \times {}^{-}\frac{4}{3} = {}^{+}\frac{4}{6} = {}^{+}\frac{2}{3}$
- $^{-}\frac{1}{2} \div {}^{+}\frac{3}{4} = {}^{-}\frac{1}{2} \times {}^{+}\frac{4}{3} = {}^{-}\frac{4}{6} = {}^{-}\frac{2}{3}$
- $^{+}\frac{1}{2} \div {}^{-}1 = {}^{-}\frac{1}{2}$
- $^{+}1.5 \div {}^{+}.3 = {}^{+}5$
- $^{+}1.5 \div {}^{-}.3 = {}^{-}5$
- $^{-}1.5 \div {}^{-}1 = {}^{+}1.5$

Commutative, Associative, and Distributive Properties of Operations on Signed Numbers

The COMMUTATIVE properties of addition and multiplication apply to signed numbers.

- $^{-}5 + {}^{+}2 = {}^{+}2 + {}^{-}5$ The terms on both sides of the equals sign are equal to $^{-}3$.

- $^{-}2 \times {}^{-}3 = {}^{-}3 \times {}^{-}2$ The terms on both sides of the equals sign are equal to $^{+}6$.

The ASSOCIATIVE properties of addition and multiplication apply to signed numbers.

- $^{-}2 + ({}^{-}3 + {}^{+}4) = ({}^{-}2 + {}^{-}3) + {}^{+}4$

  Simplifying the terms on each side of the equals sign shows that both sides are equal.

  $^{-}2 + ({}^{-}3 + {}^{+}4) = {}^{-}2 + ({}^{+}1) = {}^{-}1$

  $({}^{-}2 + {}^{-}3) + {}^{+}4 = ({}^{-}5) + {}^{+}4 = {}^{-}1$

- $^{-}2 \times ({}^{+}3 \times {}^{-}4) = ({}^{-}2 \times {}^{+}3) \times {}^{-}4$

  Simplifying the terms on each side of the equals sign shows that both sides are equal.

  $^{-}2 \times ({}^{+}3 \times {}^{-}4) = {}^{-}2 \times ({}^{-}12) = {}^{+}24$

  $({}^{-}2 \times {}^{+}3) \times {}^{-}4 = ({}^{-}6) \times {}^{-}4 = {}^{+}24$

Multiplication is DISTRIBUTIVE over addition and subtraction of signed numbers.

- $^{+}2(^{-}3 + {}^{+}4) = (^{+}2 \times {}^{-}3) + (^{+}2 \times {}^{+}4)$

  Simplifying the terms on each side of the equals sign shows that both sides are equal.

  $^{+}2(^{-}3 + {}^{+}4) = {}^{+}2\,(^{+}1) = {}^{+}2$

  $(^{+}2 \times {}^{-}3) + (^{+}2 \times {}^{+}4) = (^{-}6) + (^{+}8) = {}^{+}2$

- $^{-}3(^{+}5 - {}^{-}2) = (^{-}3 \times {}^{+}5) - (^{-}3 \times {}^{-}2)$

  Simplifying the terms on each side of the equals sign shows that both sides are equal.

  $^{-}3(^{+}5 - {}^{-}2) = {}^{-}3\,(^{+}5 + {}^{+}2) = {}^{-}3(^{+}7) = {}^{-}21$

  $(^{-}3 \times {}^{+}5) - (^{-}3 \times {}^{-}2) = (^{-}15) - (^{+}6) = {}^{-}15 + {}^{-}6 = {}^{-}21$

Order of Operations

When a complex problem includes signed numbers, use the order of operations (PEMDAS) to solve it.

- $(^{-}3)^2 - (^{+}1 + ^{-}1)^3$ — First simplify the terms in parentheses. $(^{+}1 + ^{-}1) = 0$.

  $= (^{-}3)^2 - (0)^3$ — Then find the value of the exponentials. $(^{-}3)^2 = 9$; $(0)^3 = 0$

  $= 9 - 0$ — Then subtract. $9 - 0 = 9$

  $= 9$

- $^{-}8 + ^{-}4 \times ^{+}21 - ^{+}12 \div ^{-}6 - ^{+}4$ — First multiply and divide from left to right.

  $^{-}4 \times ^{+}21 = ^{-}84$

  $^{+}12 \div ^{-}6 = ^{-}2$

  $= ^{-}8 + ^{-}84 - ^{-}2 - ^{+}4$ — Then add and subtract from left to right.

  $^{-}8 + ^{-}84 = ^{-}92$

  $= ^{-}92 - ^{-}2 - ^{+}4$ — $^{-}92 - ^{-}2 = ^{-}92 + ^{+}2 = ^{-}90$

  $= ^{-}90 - ^{+}4$ — $^{-}90 - ^{+}4 = ^{-}90 + ^{-}4 = ^{-}94$

  $= ^{-}94$

- $^{+}8 + {}^{-}4\,({}^{-}21 - {}^{+}12) \div {}^{+}6 - {}^{+}4$

First simplify the terms in parentheses.

$^{-}21 - {}^{+}12$

$= {}^{-}21 + {}^{-}12$

$= {}^{-}33$

$= {}^{+}8 + {}^{-}4\,({}^{-}33) \div {}^{+}6 - {}^{+}4$

Then multiply.

$^{-}4\,({}^{-}33) = {}^{+}132$

$= {}^{+}8 + {}^{+}132 \div {}^{+}6 - {}^{+}4$

Then divide.

$^{+}132 \div {}^{+}6$

$= {}^{+}22$

$= {}^{+}8 + {}^{+}22 - {}^{+}4$

Then add.

$^{+}8 + {}^{+}22 = {}^{+}30$

$= {}^{+}30 - {}^{+}4$

Then subtract.

$^{+}30 - {}^{+}4$

$= {}^{+}30 + {}^{-}4$

$= {}^{+}26$

$= {}^{+}26$

Exponentials with Negative Bases

Because we can multiply a negative number by itself several times, it can be written as an exponential that has a negative base.

Odd powers of negative numbers have negative values.
Even powers of negative numbers have positive values.

- $(^-2)^1 = {}^-2$
- $(^-2)^2 = (^-2)\,(^-2) = {}^+4$
- $(^-2)^3 = (^-2)\,(^-2)\,(^-2) = {}^-8$
- $(^-2)^4 = (^-2)\,(^-2)\,(^-2)\,(^-2) = {}^+16$
- $(^-2)^5 = (^-2)\,(^-2)\,(^-2)\,(^-2)\,(^-2) = {}^-32$

The procedures for simplifying exponentials with positive bases also apply to exponentials with negative bases.

- $(^-2)^7 \cdot (^-2)^2 = (^-2)^9 = {}^-512$
- $(^-2)^3 \div (^-2) = (^-2)^2 = 4$
- $(^-2)^1 = {}^-2$
- $(^-2)^0 = 1$

Exponentials with Negative Exponents

A NEGATIVE EXPONENT results when one exponential is divided by another that has the same base and a larger exponent than the one being divided.

- $\frac{2^2}{2^3} = 2^{-1}$
- $\frac{10^2}{10^5} = 10^{-3}$
- $\frac{5^1}{5^3} = 5^{-2}$

When an exponent is negative, the value of the exponential is a fraction. The numerator of the fraction is 1 and the denominator is the initial exponential with a positive exponent.

- $\frac{2^2}{2^3} = 2^{-1}$ and $\frac{2^2}{2^3} = \frac{4}{8} = \frac{1}{2} = \frac{1}{2^1}$

  $$2^{-1} = \frac{1}{2^1}$$

- $\frac{10^2}{10^5} = 10^{-3}$ and $\frac{10^2}{10^5} = \frac{100}{100000} = \frac{1}{1000} = \frac{1}{10^3}$

  $$10^{-3} = \frac{1}{10^3}$$

- $\frac{5^1}{5^3} = 5^{-2}$ and $\frac{5^1}{5^3} = \frac{5}{125} = \frac{1}{25} = \frac{1}{5^2}$

  $$5^{-2} = \frac{1}{5^2}$$

The procedures for simplifying exponentials that have positive exponents also apply to those that have negative exponents.

- $2^{-4} \cdot (2)^3 = 2^{-1}$
- $10^5 \cdot 10^0 \cdot 10^1 \cdot 10^{-4} \cdot 10 = 10^3$
- $(^-3)^{-10} \cdot (^-3)^{10} \cdot (^-3)^1 = (^-3)^1$
- $16^{-6} \div 16^0 = 16^{-6}$
- $5^{-2} \cdot \frac{5^4}{5^1} = 5^{-2} \cdot 5^3 = 5^1$
- $(8^{34})^{-1} = 8^{-34}$
- $(10^0)^{-3} = 10^0$

Multiplying and Dividing by Powers of 10 That Have Negative Exponents

Multiplying a decimal number by a power of 10 that has a negative exponent is the same as dividing the decimal by the exponential with a positive exponent.

- $.3 \times (10)^{-3} = .3 \times \frac{1}{10^3} = \frac{.3}{10^3} = .3 \div 10^3 = .0003$

$$.3 \times .001 = .001\overline{).3} = .1\overline{).300}\ \ (300)$$

To multiply a decimal number by a power of 10 that has a negative exponent, move the decimal point to the left as many places as the absolute value of the exponent of the power of 10.

- $.3 \times (10)^{-3} = .0003$

Dividing a decimal number by a power of 10 that has a negative exponent is the same as multiplying the decimal by an exponential with a positive exponent.

- $.3 \div (10)^{-3} = .3 \div \frac{1}{10^3} = .3 \times \frac{10^3}{1} = 300.$

$$.001\overline{).3} = .1\overline{).300}\ \ (300) = .3 \times 1000$$

To divide a decimal number by a power of 10 that has a negative exponent, move the decimal point to the right as many places as the absolute value of the exponent of the power of 10.

$.3 \div (10)^{-3} = 300.$

Scientific Notation (for Small Numbers)

Scientific notation is the way scientists write approximations for very small numbers as well as large ones. (See Page 89.)

Scientific notation is the product of two terms. The first term is a number between 1 and 10. The second term, for very small numbers, is a power of 10 with a negative exponent.

- 2.9×10^{-5} Find the value by moving the point 5 places to the left.

  $2.9 \times 10^{-5} = .000029$

- 7.025×10^{-10} Find the value by moving the point 10 places to the left.

  $7.025 \times 10^{-10} = .0000000007025$

Any very small number can be written in scientific notation.

Steps

1. Ignore the leading zeros and copy the remaining digits.

   .000000006031 = **6031**

2. Place a decimal point after the first digit that you copied.

   .000000006031 = **6.031**

   (6.031 is between 1 and 10.)

3. In the original number, count the digits between where the decimal point had been and where it was moved to.

   The decimal point moved over these 9 digits:

   000 000 006

4. Use the negative of this number as the exponent in the power of 10.

   .000000006031 = $\mathbf{6.031 \times 10^{-9}}$

Using Exponentials to Identify Place Values

The place values for decimal numbers are all powers of 10, and can be written using exponentials.

The ones column is in the center of the place values, just as 0 is in the center of the positive and negative numbers.

1000	100	10	1		$\frac{1}{10}$	$\frac{1}{100}$	$\frac{1}{1000}$
10^3	10^2	10^1	10^0	.	10^{-1}	10^{-2}	10^{-3}

NOTE: THE NEGATIVE EXPONENTS ARE TO THE RIGHT OF THE POSITIVE EXPONENTS BECAUSE THE PLACE VALUES GET SMALLER AS WE MOVE TO THE RIGHT.

ON A NUMBER LINE, THE NEGATIVE NUMBERS ARE TO THE LEFT OF THE POSITIVE NUMBERS, BECAUSE THE NUMBERS GET SMALLER AS WE MOVE TO THE LEFT.

Algebra
(Selected Topics)

Equation. Expression. Constant. Variable. Value of an expression. Solving an equation. Coordinate grid. x-axis. y-axis. Quadrant. (x,y) coordinates. Linear equation. Slope. Y-intercept. Quadratic equation and roots. Formula. Quadratic formula. Factorial. Combination. Permutation. Arithmetic progression. Geometric progression.

Algebraic Expressions and Equations

Algebraic expressions and equations use letters to represent numbers.

An algebraic EQUATION has an equals sign. An algebraic EXPRESSION does *not* have an equals sign.

- $5 + n$ is an expression that means "Five plus some number".
- $5 + n = 7$ is an equation that means "Five plus some number equals seven".

A number in an algebraic expression or equation is called a CONSTANT.

A letter in an algebraic expression or equation is called a VARIABLE.

- $5 + n$ — Five is a constant. n is a variable.

Algebra avoids the multiplication symbol so that it is not mistaken for the variable "x". Multiplication is either indicated by parentheses or by placing two letters or a number and a letter next to one another.

- $5x$ — Five times x.
- xy — x times y.
- $5(a + b)$ — Five times the sum of a and b.

NOTE: $x = 1x$.

Finding the Value of an Expression

To find the value of an expression, you must replace all variables with their values, and then find the result.

- To find the value of $x^2 + 2x - 7$ when $x = 4$, start by replacing each x in the expression with 4:

  $$(4)^2 + 2(4) - 7$$

  Then simplify:

  $$16 + 8 - 7 = 17$$

- To find the value of $x^2 + 2x - 7$ when $x = 10$, start by replacing each x in the expression with 10:

  $$(10)^2 + 2(10) - 7$$

  Then simplify:

  $$100 + 20 - 7 = 113$$

- To find the value of $x^2 + 2x - 7$ when $x = {}^-1$, start by replacing each x in the expression with $^-1$:

  $$({}^-1)^2 + 2({}^-1) - 7$$

  Then simplify:

  $$1 + {}^-2 - 7 = {}^-1 - 7 = {}^-1 + {}^-7 = {}^-8$$

General Guidelines for Solving All Equations

Solve an equation containing a single variable by getting the variable by itself on one side of the equals sign.

Reverse an operation to remove a term.

Keep the equation balanced by performing the same operation on both sides of the equals sign.

Check your answer by placing it in the original equation to verify that the result is true.

- $x + 3 = 24$
  $x = 21$
  Subtract 3 from both sides of the equation to get the x by itself.
  CHECK: $21 + 3 = 24$

- $24 = y - 39$
  $63 = y$
  Add 39 to both sides of the equation to get the y by itself.
  CHECK: $24 = 63 - 39$

- $4n = {}^{-}36$
  $n = {}^{-}9$
  Divide both sides of the equation by 4 to get the n by itself.
  CHECK: $4({}^{-}9) = {}^{-}36$

- $\frac{x}{12} = 6$
  $x = 72$
  Multiply both sides of the equation by 12 to get the x by itself.
  CHECK: $72 \div 12 = 6$

Guidelines for Solving Equations That Have Several Operations

Begin by moving isolated constants to the side that has no variable

First, move isolated "+" or "–" constants away from the variable.

- $2x + 1 = 27$ — Subtract 1 from both sides.
  $2x = 26$ — Divide both sides by 2
  $x = 13$ — CHECK: $2(13) + 1 = 26 + 1 = 27$

- $5n - 4 = 26$ — Add 4 to both sides.
  $5n = 30$ — Divide both sides by 5.
  $n = 6$ — CHECK: $5(6) - 4 = 30 - 4 = 26$

- $5 + \frac{x}{2} = 7$ — Subtract 5 from both sides.
  $\frac{x}{2} = 2$ — Multiply both sides by 2.
  $x = 4$ — CHECK: $5 + \frac{4}{2} = 5 + 2 = 7$

- ${}^{-}10 + \frac{x}{2} = {}^{-}4$ — Add 10 to both sides.
  $\frac{x}{2} = 6$ — Multiply both sides by 2.
  $x = 12$ — CHECK: ${}^{-}10 + \frac{12}{2} = {}^{-}10 + 6 = {}^{-}4$

Leave a positive variable in the result

Do not leave a "–" before a variable in your answer.
(If necessary, multiply both sides by –1.)

- $-x = 24$ — Multiply both sides by $^-1$.

  $x = {}^-24$

  CHECK: $-({}^-24) = 24$

- $-x + 2 = 24$ — Subtract 2 from both sides.

  $-x = 22$ — Multiply both sides by $^-1$.

  $x = {}^-22$

  CHECK: $-({}^-22) + 2 = 22 + 2 = 24$

- $4 - 8n = 28$ — Subtract 4 from both sides.

  $-8n = 24$ — Divide both sides by $^-8$.

  $n = {}^-3$

  CHECK: $4 - 8({}^-3) = 4 - {}^-24 = 4 + 24 = 28$

  NOTE: WHEN 4 WAS SUBTRACTED FROM BOTH SIDES, THE MINUS SIGN BECAME A NEGATIVE SIGN. THIS IS BECAUSE THE PROBLEM IS EQUIVALENT TO

  $4 + {}^-8n = 28$

- $-\frac{x}{2} = 4$

  Either multiply both sides by 2 first:

  $-x = 8$

  $x = {}^-8$

  Or, multiply both sides by $^-1$ first:

  $\frac{x}{2} = {}^-4$

  $x = {}^-8$

  CHECK: $-\frac{-8}{2} = -({}^-4) = 4$

Remove variables from the denominator

Remove a variable from the denominator by multiplying.

- $\frac{6}{2n} = 24$ — Multiply both sides by 2n.

  $6 = 48n$ — Divide both sides by 48.

  $\frac{6}{48} = n$ — Simplify.

  $\frac{1}{8} = n$

  CHECK: $6 \div 2(\frac{1}{8}) = 6 \div \frac{2}{8} = 6 \times \frac{8}{2} = \frac{48}{2} = 24$

- $\frac{24}{y} + 12 = 16$ — Subtract 12 from both sides.

  $\frac{24}{y} = 4$ — Multiply both sides by y.

  $24 = 4y$ — Divide both sides by 4.

  $6 = y$

  CHECK: $\frac{24}{6} + 12 = 4 + 12 = 16$

Eliminate parentheses

Eliminate parentheses by division or distribution.

- $3(z - 2) = 24$ Divide both sides by 3.

  $z - 2 = 8$ Add 2 to both sides

  $z = 10$

  CHECK: $3(10 - 2) = 3(8) = 24$

- $\frac{2(x + 5)}{5} = 24$ Multiply both sides by 5.

  $2(x + 5) = 120$ Divide both sides by 2.

  $x + 5 = 60$ Subtract 5 from both sides.

  $x = 55$

  CHECK: $\frac{2(55+5)}{5} = \frac{2(60)}{5} = \frac{120}{5} = 24$

- $5(y + 9) = 24$ Distribute the 5 into the parenthesis.

  $5y + 45 = 24$ Subtract 45 from both sides.

  $5y = {}^{-}21$ Divide both sides by 5.

  $y = \frac{-21}{5} = {}^{-}4\frac{1}{5}$

  CHECK: $5({}^{-}4\frac{1}{5} + 9) = 5(4\frac{4}{5}) = 5(\frac{24}{5}) = 24$

- $4(x - 3) = {}^{-}3$ Distribute the 4 into the parenthesis.

  $4x - 12 = {}^{-}3$ Add 12 to both sides.

  $4x = 9$ Divide both sides by 4.

  $x = \frac{9}{4} = 2\frac{1}{4}$

  CHECK: $4(2\frac{1}{4} - 3) = 4({}^{-}\frac{3}{4}) = {}^{-}3$

Combine similar terms

Get all similar terms on the same side of the equation and then combine them.

- $y + 2y - 70 = {}^{-}10$ — Combine the y terms.

  NOTE: $y = 1y$

  $3y - 70 = {}^{-}10$ — Add 70 to both sides.

  $3y = 60$ — Divide both sides by 3.

  $y = 20$

  CHECK: $20 + 2(20) - 70 = 20 + 40 - 70 = 60 - 70 = {}^{-}10$

- $2n = 3n + 5$ — Subtract 2n from both sides.

  $0 = n + 5$ — Subtract 5 from both sides.

  $-5 = n$

  CHECK: $2({}^{-}5) = 3({}^{-}5) + 5$

  ${}^{-}10 = {}^{-}15 + 5$

  ${}^{-}10 = {}^{-}10$

- $3x + 2 = 2(x - 5)$ — Distribute the 2 into the parentheses.

  $3x + 2 = 2x - 10$ — Subtract 2x from both sides.

  $x + 2 = {}^{-}10$ — Subtract 2 from both sides.

  $x = {}^{-}12$

  CHECK:

  $3({}^{-}12) + 2 = 2[({}^{-}12) - 5]$

  ${}^{-}36 + 2 = 2({}^{-}17)$

  ${}^{-}34 = {}^{-}34$

Eliminate common factors

Cancel out factors that are common to every group of terms on both sides of an equation.

- $2x + 4 = {}^-12$ — Divide both sides by 2.
  $x + 2 = {}^-6$ — Subtract 2 from both sides.
  $x = {}^-8$
  CHECK: $2({}^-8) + 4 = {}^-16 + 4 = {}^-12$

- $x^2 + x = 5x$ — Divide both sides by x.
  $x + 1 = 5$ — Subtract 1 from both sides.
  $x = 4$
  CHECK: $4^2 + 4 = 20$ and $5(4) = 20$

- $3(x + 1) = (x+1)(x - 1)$ — Divide both sides by x + 1.
  $3 = x - 1$ — Add 1 to both sides.
  $4 = x$
  CHECK: $3(4 + 1) = 3(5) = 15$
  and $(4 + 1)(4 - 1) = (5)(3) = 15$

Tip: Divide with care.

- $\frac{3x}{6} = 12$ — Divide just the 3 and the 12 by 3; not the 6.
  $\frac{x}{6} = 4$
  $x = 24$

- $2(x + 4) = 6$ — Divide just the 2 and the 6 by 2; not the 4.
  $x + 4 = 3$
  $x = {}^-1$

Solving a System of Equations

A system of equations is a group of related equations that contain information about two or more variables.

Solve a system of equations by first finding one of the variables, then substituting its value into the other equation.

- $\boxed{x + 2y = 300}$ AND $\boxed{y = 10}$

$x + 2(10) = 300$

$x + 20 = 300$

$x = 280$

Replace the y in the first equation with 10.

Then solve for x.

CHECK: $280 + 2(10) = 300$

- $\boxed{5 - 2x + y = {}^-6}$ AND $\boxed{x + 3 = 5}$

First solve for x.

$x + 3 = 5$

$x = 2$

CHECK:
$2 + 3 = 5$

Then use the value of x to solve for y.

$5 - 2x + y = {}^-6$

$5 - 2(2) + y = {}^-6$

$5 - 4 + y = {}^-6$

$1 + y = {}^-6$

$y = {}^-7$

CHECK:

$5 - 2(2) + ({}^-7)$

$= 5 - 4 + ({}^-7)$

$= 1 + ({}^-7)$

$= {}^-6$

Using Algebra to Solve Percent Problems

Every percent problem can be written as a proportion:

$$\frac{\text{Percent}}{100} = \frac{\text{Part}}{\text{Whole}}$$

where one of these three terms will be unknown.

Since equal fractions have equal cross-products, cross-multiply and solve for the unknown.

- Find 25% of 80. Find the PART.

$$\frac{25}{100} = \frac{x}{80}$$
$$25(80) = 100x$$
$$2000 = 100x$$
$$20 = x$$

- 25% of what number is 20? Find the WHOLE.

$$\frac{25}{100} = \frac{20}{x}$$
$$25x = 2000$$
$$x = 80$$

- What percent of 80 is 20? Find the PERCENT.

$$\frac{x}{100} = \frac{20}{80}$$
$$80x = 2000$$
$$x = 25$$

Multiplying Monomials and Polynomials

A MONOMIAL is a term made up of the product of several constants and variables.

- a^2y and $3x$ are monomials.

Multiply two monomials by multiplying their constants and variables together.

- $(a^2y)(3xy) = 3a^2xy^2$

A POLYNOMIAL is an expression made up of variables and constants, using the operations of addition, subtraction, as well as multiplication.

- $a^2y + 6$, $4x - 2a$, and $a^2b^2c^2 + 3c^2$ are polynomials.

Multiply two polynomials column by column, keeping similar terms in the same column.

- $$\begin{array}{r} x^2 + 3x + 7 \\ \underline{4x - 2} \\ -2x^2 - 6x - 14 \\ \underline{4x^3 + 12x^2 + 28x } \\ 4x^3 + 10x^2 + 22x - 14 \end{array}$$

Graphing on a Coordinate Grid

Graphing points on a coordinate grid

A coordinate grid is made up of squares.

The x-axis and y-axis divide the grid into four quadrants.

The intersections on the grid are identified by (x,y) coordinates. (A coordinate always begins with the x value.)

The intersection of the x and y axes has a coordinate of (0,0) and is called the ORIGIN.

Graph a point by putting a dot on the grid at its coordinates.

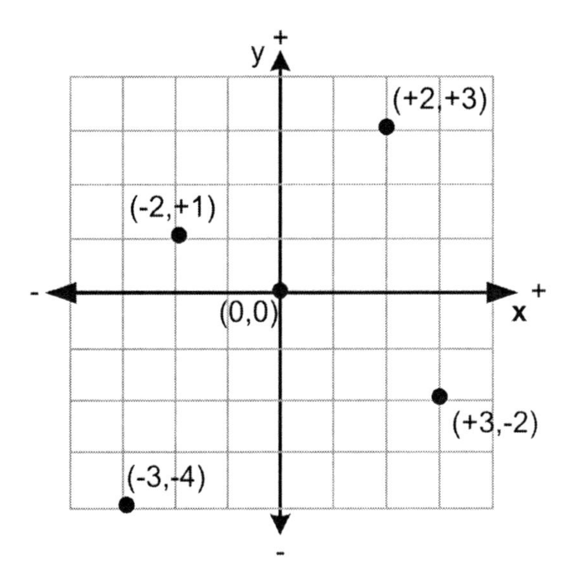

Graphing a linear equation

A LINEAR EQUATION represents the relationship of two variables that is a straight line when graphed.

A linear equation has the following form:

$$y = mx + b$$

where:

m is the SLOPE of the line, which is the change in y for each increase of x by 1
b is the Y-INTERCEPT, which is the value of y when x = 0

- $y = 2x - 4$ has a slope of 2 (y increases by two for each increase of x by 1) and a y-intercept of -4.

To graph a linear equation, select a few values for x and calculate the value of y. Then graph the points on the xy-coordinate grid and connect the dots.

- The graph of the equation $y = 2x - 4$:

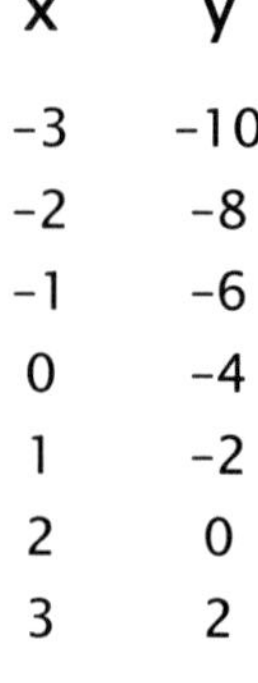

x	y
-3	-10
-2	-8
-1	-6
0	-4
1	-2
2	0
3	2

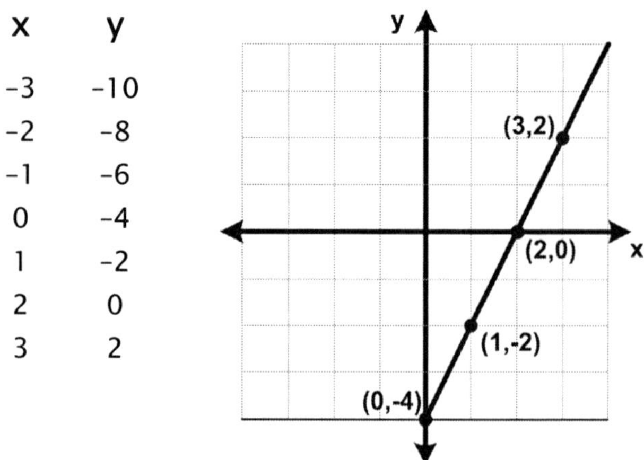

Graphing a quadratic equation

A QUADRATIC EQUATION has a cup (parabolic) shape on a graph.

A quadratic equation has the following form:

$$y = ax^2 + bx + c$$

To graph a quadratic equation, select several values for x and calculate the value of y. Then graph a few points on the xy-coordinate grid and connect the dots.

- The graph of the equation $y = x^2 - x - 2$:

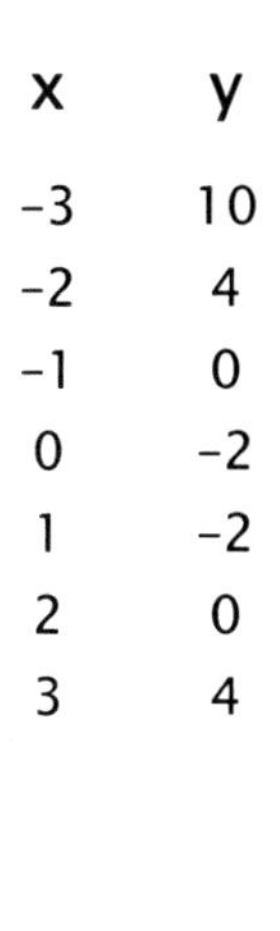

x	y
-3	10
-2	4
-1	0
0	-2
1	-2
2	0
3	4

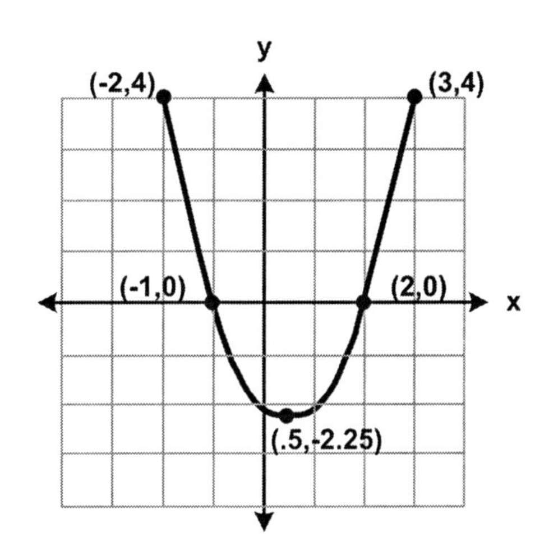

.5   -2.25

Because the minimum point must be halfway between $x = -1$ and $x = 2$, find the value of y when $x = .5$

Formulas

Formulas are algebraic equations that concisely describe a fixed relationship between stable real-world variables.

Geometry formulas

Circle formulas

Diameter	$d = 2r$
Circumference	$c = \pi d$
	$c = 2\pi r$

Perimeter formulas

Rectangle	$P = 2(l + w)$
Square	$P = 4s$
Rhombus	$P = 4s$
Equilateral Triangle	$P = 3s$

Area formulas

Rectangle	$A = lw$
Square	$A = s^2$
Parallelogram/Rhombus	$A = bh$
Triangle	$A = \frac{bh}{2}$
Circle	$A = \pi r^2$

Volume formulas

Prism	$V = lwh$
Cube	$V = s^3$

The quadratic formula

The QUADRATIC FORMULA is used to find the roots of a quadratic equation of the form:

$$y = ax^2 + bx + c$$

The ROOTS are the values of x when a parabola intersects the x-axis (y = 0) of a coordinate grid.
(Refer to the graph on Page 175.)

The quadratic formula is:

$$x = \frac{-b \pm \sqrt{b^2 - 4ac}}{2a}$$

where $\pm$ means "plus or minus".

- To solve for the two values of x in the equation:

  $$y = x^2 - x - 2$$

  begin by identifying the values for a, b, and c.

  $$a = 1 \quad b = -1 \quad c = -2$$

  Substitute these values into the quadratic formula.

  $$x = \frac{-(-1) \pm \sqrt{(-1)^2 - 4(1)(-2)}}{2(1)}$$

  Simplify the formula:

  $$x = \frac{1 \pm \sqrt{(-1)^2 - (-8)}}{2} = \frac{1 \pm \sqrt{1+8}}{2} = \frac{1 \pm \sqrt{9}}{2} = \frac{1 \pm 3}{2}$$

  Solve for the two values of x:

  $$x = \frac{1+3}{2} = \frac{4}{2} = 2 \quad \text{and} \quad x = \frac{1-3}{2} = \frac{-2}{2} = -1$$

CHECK:

Substitute these values of x into the original formula to verify that $y = 0$.

When $x = 2$,

$y = x^2 - x - 2 = 4 - 2 - 2 = 0$

When $x = {}^{-}1$,

$y = x^2 - x - 2 = 1 - ({}^{-}1) - 2$
$= 1 + 1 - 2 = 0$

The coordinates of these points are (-1,0) and (2,0).

Factorials

The FACTORIAL of a number is the product of that number, times one less, times one less, ... until 1 is reached.

- 6 factorial equals $6 \times 5 \times 4 \times 3 \times 2 \times 1 = 720$
- 3 factorial equals $3 \times 2 \times 1 = 6$

The symbol for factorial is the exclamation point (!).

$$n! = (n)(n-1)(n-2)(n-3) \ldots (2)(1)$$

- 6 factorial is written as 6!
  $6! = 6 \times 5 \times 4 \times 3 \times 2 \times 1 = 720$
- 3 factorial is written as 3!
  $3! = 3 \times 2 \times 1 = 6$

Zero factorial is defined to be 1.

$0! = 1$

Combinations

A COMBINATION is the number of all possible distinct groups that can be made from a number of items.

The formula used to calculate a combination is:

$$_nC_r = \frac{n!}{r!(n-r)!}$$

where:

n is the total number of items
r is the number of items in a group

Of the four letters A, B, C, and D:

- $_4C_4 = 4!/4!(4 - 4)! = 4!/4!(0!) = 24/24(1) = 24/24 = 1$

  ABCD

- $_4C_3 = 4!/3!(4 - 3)! = 4!/3!(1!) = 24/6(1) = 24/6 = 4$

  ABC ABD ACD BCD

- $_4C_2 = 4!/2!(4 - 2)! = 4!/2!(2!) = 24/2(2) = 24/4 = 6$

  AB AC AD BC BD CD

- $_4C_1 = 4!/1!(4 - 1)! = 4!/1!(3!) = 24/1(6) = 24/6 = 4$

  A B C D

Permutations

A PERMUTATION is the number of all possible ordered arrangements that can be made from a group of items. The formula used to calculate a permutation is:

$$_nP_r = \frac{n!}{(n-r)!}$$

where:

n is the total number of items
r is the number of items in an arrangement

Of the four letters A, B, C, and D:

- $_4P_4 = 4!/(4 - 4)! = 4!/(0)! = 24/1 = 24$

ABCD	BACD	CABD	DABC
ABDC	BADC	CADB	DACB
ACBD	BCAD	CBAC	DBAC
ACDB	BCDA	CBAD	DBCA
ADBC	BDAC	CDAB	DCAB
ADCB	BDCA	CDBA	DCBA

- $_4P_3 = 4!/(4 - 3)! = 4!/(1)! = 24/1 = 24$

ABC	BAC	CAB	DAB
ABD	BAD	CAD	DAC
ACB	BCA	CBA	DBA
ACD	BCD	CBA	DBC
ADB	BDA	CDA	DCA
ADC	BDC	CDB	DCB

- $_4P_2 = 4!/(4 - 2)! = 4!/(2)! = 24/2 = 12$

AB	BA	CA	DA
AC	BC	CB	DB
AD	BD	CD	DC

- $_4P_1 = 4!/(4 - 1)! = 4!/(3)! = 24/6 = 4$

A	B	C	D

Arithmetic progressions

An ARITHMETIC (ar-ith-met-ik) PROGRESSION is a sequence of numbers that increase by repeatedly adding a constant to the last number.

- {7, 11, 15, 19, ...} is an arithmetic progression where each number is four more than the previous number.

The formula used to find the nth number in the sequence $a_1, a_2, a_3, \ldots, a_n$ is:

$$a_n = a_1 + (n - 1)d$$

where:

n is the position of the nth number

a_n is the nth number

a_1 is the first number

d is the common difference between the numbers

- Use this formula to find the 10th number in the sequence: {7, 11, 15, 19, ...}.

  $a_1 = 7$, $n = 10$, and $d = 4$

  $a_{10} = 7 + (10 - 1)(4) = 7 + (9)(4) = 7 + 36 = 43$

The formula used to find the sum of the first n numbers is:

$$S = \frac{n[2a_1 + (n-1)d]}{2}$$

- Use this formula to find the sum of the first three numbers in the sequence: {7, 11, 15, 19, ...}.

  $a_1 = 7$, $n = 3$, and $d = 4$

  $S = 3[2(7) + (3 - 1)(4)] / 2 = 3[2(7) + (2)(4)] / 2$

  $= 3[14 + 8]/2 = 3[22]/2 = 66/2 = 33$

Geometric progressions

A GEOMETRIC PROGRESSION is a sequence of numbers that increase by repeatedly multiplying the last number by a constant.

- {3, 6, 12, 24, ...} is a geometric progression where each number is two times the previous number.

The formula used to find the nth number in the sequence $a_1, a_2, a_3, \ldots, a_n$ is:

$$a_n = a_1(r^{n-1})$$

where:

n is the position of the nth number

a_n is the nth number

a_1 is the first number

r is the multiple by which the numbers increase

- Use this formula to find the 6th number in the sequence: {3, 6, 12, 24, ...}.

  $a_1 = 3$, $n = 6$, and $r = 2$

  $$a_6 = 3(2^{6-1}) = 3(2^5) = 3(32) = 96$$

Index

A

Absolute value **132**
Acute angle **118**
Adding
- Decimal numbers 68
- Fractions 40
- Mixed Numbers 41
- Positive & negative numbers 140
- Whole numbers 5

Algebra **161**
Angle **117**
Approximately **20**
Area **126**
- Formulas 176

Arithmetic progression **181**
Ascending order **39**
Associative property **98**
- Positive & negative numbers 151

Average **114**

B

Base of an exponential **85**
- Negative 155

Borrowing when subtracting
- Fractions 49
- Whole numbers 11

C

Canceling in fraction
- Division 59
- Multiplication 56

Carrying when adding whole numbers **8**
Centimeters **124**
Checking your answer
- Dividing whole numbers 18
- Subtracting whole numbers 11

Circle
- π 123
- Area 127
- Circumference, diameter, radius 122
- Formulas 176

Circumference **122**
Combinations **179**
Commas used in numbers **2**
Common denominator **28**
- Least 38
- When adding fractions 42
- When ordering fractions 39
- When subtracting fractions 48

Common factor **25**
Common multiple **23**
Commutative property **97**
- Positive & negative numbers 151

Complex fractions **61**
Composite number **26**
Constant **161**
Converting
- Decimal to and from percent 107
- Decimal to fraction 75
- Fraction to and from percent 108
- Fraction to decimal number 74
- Fraction to fraction in higher terms 34

Fraction to fraction in lowest terms 33
Improper fraction to mixed or whole number 35
Mixed number to decimal number 75
Mixed number to improper fraction 37
Repeating decimal number to fraction 76
Whole number to improper fraction 36

Coordinate grid **173**

Counting numbers **1**

Cross-products **32**

Cube
Volume 129

Cubes and cube roots **82**

D

Data set **114**

Decimal numbers **63**
Adding 68
Dividing 72
Fraction conversion 74
Multiplying 70
Non-terminating 67
Ordering 66
Reading 64
Repeating 67
Rounding 65
Subtracting 69
Terminating 67

Degree **117**

Denominator **28**

Descending order **39**

Diameter **122**

Difference **9**

Digits **1**

Distributive property **99**
Positive & negative numbers 152

Dividend **17**

Dividing
By powers of 10 84
By zero 18
Decimal numbers 72
Fractions 58
Long division 20
Mixed numbers 58
Positive & negative numbers 149
Short division 19
Whole numbers 16

Divisibility rules **21**

Divisor **17**

E

Empty place values **2**

Equal fractions **31**

Equation **161**
Linear 174
Quadratic 175

Equilateral triangle **121**
Perimeter 125

Even number **21**

Expanded form **2**

Exponent **85**
Fractional 91
Negative 156
No exponent 85
One 85
Zero 90

Exponential notation **85**
Powers of fractions and decimal numbers 92
Prime factorization 88
Simplifying 86

Expression, algebraic **161**
Value of 162

F

Factor **24**

Factor tree **27**

Factorial 178
Feet 124
Formulas 176
  Area 176
  Arithmetic progression 181
  Circle 176
  Combination 179
  Factorial 178
  Geometric progression 182
  Perimeter 176
  Permutation 180
  Quadratic 177
  Volume 176
Fractions 28
  Adding 40
  As division 29
  Comparing operations 62
  Complex 61
  Decimal conversion 74
  Dividing 58
  Equivalence 31
  Higher terms 34
  Improper fraction 30
  Lowest terms 33
  Multiplying 52
  Ordering 39
  Proper fraction 30
  Reciprocal 57
  Reducing 33
  Subtracting 46

G

Geometric progression 182
Geometry 117
Graphing
  On a coordinate grid 173
  On a number line 102
Greater than symbol 101
Greater than or equal to symbol 101
Greatest Common Factor (GCF) 25

H

Higher terms 34
Hypotenuse of a right triangle 121
  And perimeter 125

I

Improper fraction 30
  Simplest form 35
Inches 124
Inequality symbols 101
Integers 136
Irrational numbers 137
Irregular polygon 119
Isosceles triangle 121

K

Kilometers 124

L

Least common denominator (LCD) 38
Least common multiple (LCM) 23
Length 124
Less than or equal to symbol 101
Less than symbol 101
Line segment 117
Linear equation 174
Lines
  Parallel lines 119
  Perpendicular lines 119
Long division 20
Lowest terms 33

M

Mean 114

Measurement
- Area 126
- Length 124
- Perimeter 125
- Surface area 128
- Volume 129

Median **115**

Meters **124**

Metric units
- Area 126
- Length 124

Miles **124**

Millimeters **124**

Mixed number **30**
- Adding 41
- Dividing 58
- Multiplying 54
- Subtracting 47

Mode **114**

Monomial **172**

Multiples **22**

Multiplying
- By powers of 10 83
- Decimal numbers 70
- Fractions 52
- Mixed numbers 54
- Multiplication table 12
- Positive & negative numbers 146
- Whole numbers 12

N

Natural numbers **1**

Negative numbers **130**
- Exponentials 155

Non-repeating decimal number **67**

Non-terminating decimal number **67**

Notation
- Exponential notation 85
- Place-value notation 2
- Scientific notation
  - Very large numbers 89
  - Very small numbers 159
- Set notation 77

Number line **102**
- Positive & negative numbers 131

Numerals **1**

Numerator **28**

O

Obtuse angle **118**

Odd number **21**

Operations **4**

Operators **4**

Opposites **134**

Order of operations **94**
- Positive & negative numbers 153

Ordering
- Decimal numbers 66
- Fractions 39
- Fractions & decimal numbers 76
- Positive & negative numbers 138

Origin
- Coordinate grid 173
- Number line 131

P

Parallel lines **119**

Parallelogram **120**
- Area 127

PEMDAS **95**

Percent **106**
- And algebra 171
- Converting decimals to and from 107
- Converting fractions to and from 108

Greater than 100 112
Increase or decrease 111
Problems 109
Perimeter **125**
Formulas 176
Permutations **180**
Perpendicular lines **119**
Pi (π) **123**
Placeholder (0) **2**
Place-value notation **2**
Decimal numbers 63
Using exponential notation 160
Plane **119**
Polygon **119**
Polynomial **172**
Positive and negative numbers **130**
Adding 140
Dividing 149
Multiplying 146
Negative exponentials 155
Opposites 134
Order of operations 153
Ordering 138
Subtracting 142
Types 136
Power **78**
Number of 79
Of a power 87
Powers of 10 **79**
As place values 160
Dividing by 84
Exponential notation 88
Multiplying by 83
Negative exponent 158
Powers of negative numbers **155**
Prime factor tree **27**
Prime factorization **27**
And exponential notation 88
Prime numbers **26**
Prism
Surface area 128
Volume 129
Probability **113**
Product **13**
Progression
Arithmetic 181
Geometric 182
Proper fraction **30**
Properties of operations **97**
On signed numbers 151
Proportion **105**
Pythagorean Theorem **121**
And perimeter 125

Q

Quadrant **173**
Quadratic
Equation 175
Formula 177
Quadrilateral **120**
Quotient **17**

R

Radical **80**
Radius **122**
Range **116**
Ratio **104**
Rational numbers **136**
Real numbers **137**
Reciprocals **57**
Rectangle **120**
Area 127
Perimeter 125
Reducing a fraction to lowest terms **33**
Regular polygon **119**

Relatively prime numbers **26**

Remainder **18**
  As fraction 29

Repeated addition **12**

Repeating decimal number **67**

Rhombus **120**
  Area 127
  Perimeter 125

Right angle **118**

Right triangle **121**

Roots
  Cube 82
  Square 80

Rounding
  Decimal numbers 65
  Whole numbers 3

S

Scalene triangle **121**

Scientific notation
  For very large numbers 89
  For very small numbers 159

Series of numbers
  Arithmetic progression 181
  Geometric progression 182

Set notation **77**

Short division **19**

Sign
  Positive & negative 131

Signed numbers **132**
  See also: Positive and negative numbers

Significant digit **65**

Simplest form
  Of a fraction 33
  Of an improper fraction 35

Slope of a line **174**

Solving equations
  General guidelines 163
  Having several operations 164

Square **120**
  Area of 127
  Perimeter 125
  Units 126

Square root calculation **81**

Squares and square roots **80**

Statistical measures **114**

Straight angle **118**

Subtracting
  Decimal numbers 69
  Fractions 46
  Mixed Numbers 47
  Positive & negative numbers 142
  Whole numbers 9

Sum **6**

Surface area **128**

Symbols
  Absolute value | | 132
  Approximately ≈ 20
  Cube root $\sqrt[3]{\ }$ 82
  Factorial ! 178
  Inequality ≠, <, >, ≤, ≥ 101
  Operators +, –, x, ÷ 4
  Percent % 106
  Pi π 123
  Plus or minus ± 177
  Positive & negative signs +, – 131
  Set { } 77
  Square root (radical) $\sqrt{\ }$ 80

System of equations **170**

T

Terminating decimal number **67**

Trapezoid **120**

Triangle: equilateral, isosceles, right, scalene **121**

U

Unequal to symbol **101**

Units of measurement

- Area 126
- Length 124

V

Value of an expression **162**

Variable **161**

Volume **129**

- Formulas 176

W

Whole numbers **1**

- Adding 5
- Dividing 16
- Multiplying 12
- Rounding 3
- Subtracting 9

X

X-axis **173**

Y

Y-axis **173**

Y-intercept **174**

Yards **124**

Z

Zero **1**

- Dividing by 18
- Exponent 90
- Factorial 178

LaVergne, TN USA
27 April 2010
180710LV00004B/160/P

9 780615 273907